BEING AND BIOLOGY

Other books by Brenda Dunne and Robert Jahn

Consciousness and the Source of Reality

Filters and Reflections

Margins of Reality

Molecular Memories

Quirks of the Quantum Mind

BEING AND BIOLOGY

Is Consciousness the Life Force?

Edited by Brenda Dunne and Robert Jahn

ICRL Press
Princeton, New Jersey

Table of Contents

Introduction

The greatest mystery confronting human consciousness is itself. Notwithstanding millennia of questioning, we still cannot even agree on just what consciousness is, although we acknowledge that it exists, if for no other reason than that we are capable of asking the question. To paraphrase Descartes, we know we are alive because we have consciousness.

Most contemporary science focuses its attention on the attributes of physical phenomena and tends to assume that consciousness is an emergent property of brain activity. Yet there is a vast body of evidence, both scholarly and experiential, that refutes this view and is inconsistent with the so-called laws of established physics and contemporary neuroscience.

This volume offers a compendium of empirical and theoretical perspectives from a broad range of scholarly research and personal experiences that contradict these views. What emerges is the suggestion that what we regard as consciousness may emerge from a far more profound source—the Life Force, or what Henri Bergson termed the *élan vital*, the animating principle that underlies and unifies mind, body, and spirit in all living things. A similar concept can be found in the Hindu Vedānta, which promotes the idea that the subjective evolution of consciousness is the developing principle of the world.

The opening chapter, "Is Consciousness the Life Force?", reviews some of the biological implications of the research in anomalous human/machine interactions and remote perception, conducted at the Princeton Engineering Anomalies Research (PEAR) laboratory between 1979 and 2007, which raised the topic addressed in this volume. The question is then addressed by Richard Blasband within the context of Wilhelm Reich's extensive explorations of "Orgone Energy and the Life Force," followed by Antonio Giuditta's eloquent overview of the phylogenetic origin of the human mind and brain and its implications for the nature of consciousness.

Rupert Sheldrake's reflections on "Evolution and Formative Causation" offer a comprehensive comparison between mechanistic and vitalistic approaches to the understanding of life. His model of morphogenetic fields proposes that these fields play a causal role in the development and maintenance of the forms of systems at all levels of complexity. Another provocative hypothesis is suggested by Ulisse Di Corpo and Antonella Vaninni in "A Syntropic Model of Consciousness," based on ideas originally put forward by mathematician Luigi Fantappiè, who suggested that consciousness is a property of a symmetric and complementary energy to the force of entropy, which flows from the past, to a force called syntropy, which propagates from the future, in a dynamic interplay that sustains life.

Wagner Alegretti then describes the results of some of his preliminary experiments in "Bioenergy Detection via fMRI," wherein he was able to detect the presence of subtle bioenergy outside the physical skull while subjects consciously attempted to alter their vibrational states. Some of the most exciting research in the area of alternative healing have been carried out over 35 years by William Bengston in numerous *in vivo* and *in vitro* experiments that have produced the first successful full cures of transplanted mammary cancer and methylcholanthrene-induced sarcomas in mice, and on human breast cancer cells. In "Some Speculations on Consciousness" he proposes that healing is not fundamentally a matter of "energy," but of "information."

The role of light and the pineal gland in living systems is discussed in Brennan Kersgaard's essay on "Light, Biology and Consciousness," and that of water is explored in Nelson Abreu's "Qi-Water Bridge as a Bilateral Consciousness-Brain Interface." Water memory is a fundamental principle of homeopathy, a long-established healing system that still remains disputed by mainstream medicine. In "The Physical Nature of the Biological Signal," Yolène Thomas, a colleague of the late Jacques Benveniste, describes some of the evidence that supports the efficacy of homeopathic treatment.

Living systems can not only acquire information from light and water, they also do so via sound. Thomas Anderson demonstrates how sound vibrations influence biological organisms in "The Unifying Principles of Sound." The role of the heart in intuition is a topic that has been studied extensively at the Heartmath Institute in Colorado. Rollin McCraty, the Institute's Director of Research, describes some of the studies that have explored and substantiated the importance of the heart in "Intuitive Intelligence."

Complementing the scholarly research that provides evidence for the union of mind, body, and spirit in living systems, Steve Curtis shares the personal journey that brought him from a diagnosis of a terminal cancer to full health.

The essay by Larry Dossey, "Brains and Beyond," offers an overview of the prevailing models of health and healing, highlighting their shortcomings and indicating the need for more holistic approaches that recognize empathy and compassion as vital aspects of health and healing, and biological connectivity as a natural expression of consciousness that helps sustain life as we know it.

The concluding chapter by Vasileios Basios, "Complexity, Complementarity, Consciousness," explores the role of complementarity and complexity in logic, neurosciences, psychology, and philosophy. He notes that these all point to the need for a new kind of understanding and a new kind of science—a "science towards the origins"—with consciousness as its origin.

The so-called "mind/body problem" that has plagued scientists and philosophers since the time of Plato is based on the idea that all the laws of nature are deterministic and that the world is causally closed. In this view, every physical event can be reduced to the microscopic motions of physical particles, and so consciousness must therefore be the product of events determined by the activity of the brain's atoms and molecules and there is no way that an immaterial mind can cause something physical to happen without recourse to an underlying physical derivation. Thus, the basis of the "problem."

Yet, upon personal reflection (a clearly subjective activity) and consideration of the evidence presented in these essays, it becomes evident that this deterministic dualistic perspective is flawed. All living systems evolve and are maintained by the reciprocal exchange of information with their environments. Life is comprised of both mind and matter, and is thus dependent upon this symbiotic dynamic. The division of these two dimensions of experience, and the possible existence of one without the other is the only real "problem."

If consciousness and the environment are complementary perspectives of a unitary system, then the whole is reflected in all the parts and all the parts are connected. The evidence presented in the essays in this book suggests that there is an unbroken, nonlocal, collective aspect of consciousness that links distant individuals and events—a kind of resonant connectedness that defies separation in space and time. The implications of this representation of reality are immense, altering our understanding of health and healing; of our relationships with each other, Nature, and the Cosmos; and perhaps most importantly, of our view of ourselves.

As David Bohm observed, "Deep down the consciousness of mankind is one."

Is Consciousness the Life Force?

BRENDA DUNNE AND ROBERT JAHN

Scholarly attention to the study of anomalous consciousness-related phenomena has tended to be approached from one of four perspectives: psychology; physics; the paranormal; or skeptical denial. Nearly three decades of research at the Princeton Engineering Anomalies Research (PEAR) laboratory provides reason to believe that a fifth perspective, namely biology, could contribute valuable insights in attempts to understand these events.

Consciousness itself is one of the fuzziest of conceptual categories, and for much of the 20th century even the word itself was seldom invoked in most academic circles. More recently, many mainstream scientists have acknowledged that the subjective dimensions of experience can no longer be ignored, although most of them approach such study with their reductionist spectacles firmly in place. Moving out from familiar categories and deploying familiar research strategies, scientists from various disciplines have sought a physical basis for consciousness, which they confidently presume is tucked away somewhere in the recesses of the brain.

Have we been asking the wrong questions? Is the assumption that objective physical reality is independent of subjective mental reality, which is no more than a derivative of our brain activity, incorrect? Is there reason to believe that this duality might itself be the product of consciousness rather than an attribute of the physical domain? And is it possible that consciousness is inherently linked to the biological processes of life itself?

Consciousness-related anomalies, also known as psychic or psi phenomena, have been studied scientifically since at least 1882, when the Society for Psychical Research was established in London by some of the most prominent scholars of the day. Although more than a century of rigorous research has established the reality of these effects beyond question, their source and mechanism continue to elude understanding and a comprehensive model is still lacking. This is particularly problematical for researchers because these events fail to fit neatly into any contemporary conceptual categories and suggest that something we know "for certain" could possibly be wrong—a potential source of cognitive dissonance for people committed to a search for certainty. While most of the efforts to explain consciousness-related anomalies have tended to approach the problem from the perspectives of either physics or psychology, the empirical evidence is often inconsistent with the prevailing materialist conceptions that underlie both of these fields and suggests that the phenomena may be more closely associated with a different domain altogether, namely the biology of living systems. They even may be evidence of the "life force" that Henri Bergson termed *élan vital.*

Evidence that subjective factors can tangibly influence objective physical processes indicates that the worlds of mind and matter intersect, although often in inexplicable ways. Experiments at the PEAR laboratory conducted between 1979 and 2007 in Princeton University's School of Engineering and Applied Science, and related research elsewhere, shed some light, albeit quite dim, on those uncertain areas where prevailing models of reality break down, where the division between mind and matter blurs, and where phenomena arise that don't fully belong in either. If useful insights in those areas are to be achieved, the anomalous events that emerge from the cracks between the worlds must take both worlds into account.

The primary purpose of the PEAR research program was to explore the role of consciousness in the establishment of physical reality via systematic study of engineering-related anomalies arising in human/machine interactions and controlled experiments in remote perception. The former typically consisted of changes in the output of various random physical systems that were correlated with the pre-stated intentions of their human participants. The latter explored the ability of untrained volunteers to describe geographical locations remote in distance and time, and the development of analytical methods to quantify the results thus obtained. Both of these research categories have been described in detail in a number of references,[1][2][3][4][5] so here we will simply summarize briefly those results that seem relevant to the present thesis.

- Small shifts in the means of these output distributions were consistently observed in the human/machine experiments. No single event could be identified as anomalous in itself, but over the

course of many millions of trials, the accumulation of these minute effects has been shown to be statistically significant and not
attributable to chance;

- These effects were observed across a broad array of random
physical devices and processes, ranging in scale and physical
source from a microscopic electronic random event generator to a
macroscopic random mechanical cascade;

- Experiments in remote perception have indicated that untrained
individuals are able to acquire information about spatially and
temporally remote geographical locations with no recourse to any
known sensory channels. The primary focus of this portion of the
PEAR program was the development of analytical techniques for
quantifying the results. Many such techniques were explored and
refined, yet, as the scoring methods became more sophisticated
and complex, the anomalous effects deteriorated. Nonetheless, a
database of over 650 trials displayed composite results that were
statistically significant with a probability of occurring by chance
of 3 parts in 100,000,000 (3×10^{-8});

- A body of experiments in remote human/machine interactions
also produced statistically significant effects that displayed no attenuation due to increased separations of distance or time, up to
thousands of miles and several days;

- Gender was found to be a significant factor across all the PEAR
experiments, with males and females typically producing distinctly
different overall patterns of results. Even the baselines, or "control"
conditions, where no stated intention is involved, display anomalous gender-specific shifts of the distribution means. The male
baselines tend to show reduced variances and means that remain
very close to chance, while the female baselines tend to display
large variances and means that depart significantly from chance;

- Experiments with two operators working together with a shared
intention, referred to as "co-operators," produce effects that are
on average twice as strong as those produced by the same individuals working alone, *as long as they are of opposite sex.* The
output of same-sex pairs showed no effects beyond chance, even
if the participants previously had been individually successful.

If the opposite-sex co-operators were emotionally bonded couples however, their average effects were nearly seven times larger
than their individual results.

• Over the course of repeated experiments produced by a given
operator, the first series repeatedly indicated strongest effect—a
kind of "beginner's luck"—a replicable pattern that was well beyond chance expectation. In this context, it is worth noting that
these initial efforts entailed the highest degree of uncertainty in
the operators' minds, when they had no expectations of the eventual outcomes. The second series, on the other hand, where the
participants usually had expectations based on their prior experience, displayed overall trends that were significantly *opposite*
to intention to a degree comparable to that of the successful first
series. The desired effects reappeared in later series and usually
tended to stabilize, but to a degree somewhat less than those produced in the first series.

This empirical evidence cannot be accommodated within the prevailing laws of physics and is inconsistent with most of what is known in contemporary cognitive science, but it displays numerous correlations with
biological processes associated with adaptation and evolution.

In addition to the above quantitatively demonstrated results, several
informative anecdotal observations were also acquired. For example, participants frequently reported that they felt they achieved better results
when they weren't consciously trying—a playful attitude and friendly environment seemed more effective than focused attention. They also emphasized that holding an intention was less important than establishing an
emotional resonance with the task. Indeed, they often noted that excessive cognitive activity frequently resulted in reduced or even null effects.
These subjective reports, together with the objective quantitative results,
all point to phenomena that are more closely related to a non-physical and
non-psychological connection between consciousness and the physical
world that may have biological implications.

A number of indicative studies conducted by Dean Radin and others,
reported in Radin's book *Supernormal*[6] also provide evidence for a biological correlation with psi abilities. For example, he describes experiments

in presentiment, or precognition, where such physiological processes as skin conductance, pupil dilation, and brain activity displayed significant effects prior to the presentation of stimuli.

Another indication that such anomalies are not driven by superficial cognitive processes is reflected in animal studies such as those reported by René Peoc'h, who arranged for newly hatched chicks to imprint on a randomly driven robot.[7] When the chicks were separated from their "mother" by being confined in a cage at one end of the robot's range of activity, the robot was observed to spend a significantly disproportionate amount of time in the vicinity of the chicks' cage. Such ability of baby chickens to affect the behavior of a random process with which they were emotionally bonded lends further credence to the importance of emotional bonding and the irrelevance of neo-cortical cognitive activity in generating these effects. And, of course, the unanticipated gender correlations in the PEAR experiments comprise yet another indication for a biological basis for consciousness-related anomalies.

The widespread phenomenon of dowsing, where individuals employ rods or pendulums to detect underground water, natural resources, or other hidden objects has been attributed to cryptesthesia, whereby the practitioner makes unconscious observations of the terrain and involuntarily influences the movement of the rod or pendulum. It has been suggested that people's subconscious minds may influence their bodies without conscious effort.[8]

It is well known that many biological processes such as breathing, cardiac activity, endocrine functions, etc. operate at an unconscious level, and that our emotional experiences affect our physiology as well as our minds. It thus may be that these anomalous phenomena also are expressions of a deep unconscious biological organizing capacity. Indeed, the most parsimonious representation of the empirical laboratory effects is that somehow the consciousness of the human operator has ever so slightly altered the inherent 50/50 probability of various random binary processes in the desired directions. This concept has been developed more fully in a conceptual model titled "A Modular Model of Mind/ Matter Manifestation (M5)."[9] In this representation, the distinction between mental and physical reality becomes weaker as one descends more deeply into the domains of the unconscious mind and intangible physical world, which ultimately merge into each other and into the "primordial Oneness" or "Source." Such a process is metaphorically akin to the descent of the mythological Psyche into the underworld of the unconscious in her desire to be reunited with Love (or what might be referred to in scientific parlance as "resonance").

If consciousness-related anomalies indeed prove to arise from an unconscious biological source, one would assume that such a correlation would serve a purpose advantageous to the development or survival of the organism. Recall that we are dealing here with evidence that subjective consciousness can affect quantifiable random processes in the physical world, phenomena that lie at the heart of the so-called "mind/body" paradox. The empirical effects are very small and probabilistic in nature, manifesting only collectively as measureable shifts in the means of otherwise random distributions, but collectively display significant correlations with intention (perhaps unconsciously), are gender-related, and appear to be enhanced by emotional resonance. Perhaps most striking is the evidence that these effects are independent of distance and time, thus defying the attenuation that would be evident in any known physical transmission of energy or information.

The relationship between mind (consciousness) and body (matter) has challenged philosophers since the time of Plato, yet it remains a contentious issue to this day, where it is commonly referred to as the "mind/body problem." The 17th century philosopher René Descartes contemplated the mystery of the mind/body dichotomy extensively, and finally concluded that mind was distinct from matter, but could nevertheless influence it. He was unable to explain how such interactions were actually produced, however, and hypothesized that the mind exerted control over the brain via the pineal gland, which he regarded as "the seat of the soul."[10] His heroic, but less than satisfying, attempt at an explanation, described in his *Passions of the Soul*, was that

> [the] mechanism of our body is so constructed that simply by this
> gland's being moved in any way by the soul or by any other cause,
> it drives the surrounding spirits towards the pores of the brain,
> which direct them through the nerves to the muscles; and in this
> way the gland makes the spirits move the limbs.[11]

The scientific, philosophical, and religious representations of the world that have guided Western culture over subsequent centuries have all adhered to the dualistic Cartesian perspective that leaves little space for uncertainty, and have become so habitual as to be virtually unconscious. Yet, this inherent dichotomy has determined how we categorize our experiences in order to represent them to ourselves and others. And, of course, it determines what kinds of questions we ask and how we interpret the responses we get. In this way, we have been able to construct relatively comfortable and stable intellectual structures and alleviate much psychological discomfort. In *The Doors of Perception*, Aldous Huxley observed:

> To make biological survival possible, Mind at Large has to be funneled through the reducing valve of the brain and nervous system. What comes out at the other end is a measly trickle of the kind of consciousness which will help us stay alive on the surface of this particular planet. To formulate and express the contents of this reduced awareness, man has invented and endlessly elaborated those symbol-systems and implicit philosophies which we call languages. Every individual is at once the beneficiary and the victim of the linguistic tradition into which he or she has been born—the beneficiary inasmuch as language gives access to the accumulated records of other people's experience, the victim in so far as it be-devils his sense of reality, so that he is all too apt to take his concepts for data, his words for actual things.[12]

Nonetheless, Descartes' famous *"cogito ergo sum"* implies that he believed the soul, or mind, to be the primary agent. That is, that what we now term "consciousness" is the agent of thought rather than the product of it.

It is useful to keep in mind that in studying consciousness, science is compelled to study itself, for science ultimately is an activity of consciousness. Over its several centuries of activity it has divided itself, amoeba-like, into smaller and more specialized sub-sciences. But before we can achieve the eventual unification of science it will be necessary to recognize that each of these is no more than a conceptual category, a reflection of one facet of the consciousness that is the organizing principle of all knowledge.

Despite its ubiquity, the nature of consciousness remains unknown and even its definition is unclear. Most often it is associated with the activity of the physical brain and is regarded as the source of all experiential psychological phenomena such as wakefulness, sensation, intellect, cognition, or attention. An alternative, albeit no less imprecise, characterization was first proposed in our book *Margins of Reality*[13] where we portrayed consciousness as subsuming "all categories of human experience, including perception, cognition, intuition, instinct, and emotion, at all levels, including those commonly termed 'conscious,' 'subconscious,' 'superconscious,' or 'unconscious,' without presumption of specific psychological or physiological mechanisms." We spoke of its being engaged in an ongoing exchange of information with its physical environment, defined as including "all circumstances and influences affecting the consciousness that it perceives to be separate from itself, including, as appropriate, its own physical corpus and physical habitat, as well as all the intangible psychological, social, and historical influences that bear upon it." In this view, the interface between consciousness and environment is regarded as subjective and situation-specific and it is in the process of this exchange of information that reality is constituted. This definition is consistent with William James' observation that consciousness creates its reality by selectively sorting out from the "blooming, buzzing, confusion" of environmental stimuli those portions pragmatically useful to its purposes or otherwise reinforcing the expression of its will.[14] While this characterization remains admittedly imprecise, it has the advantage of being relatively unencumbered by mechanistic assumptions, and leaves room for the self to reveal its experiences on its own terms.

In seeking an unconscious biological source that might provide an advantage for such consciousness-related anomalies in the development or survival of a biological organism, it is useful to recall that we are dealing here with evidence that subjective consciousness can affect random processes in the objective physical world. On deeper reflection, it becomes evident that there are random components in virtually all physical phenomena, and here we come to the heart of the so-called "mind/body" paradox.

It is not difficult to identify circumstances in the biological domain where a minute increase of information or order can produce a substantial effect. On the microscopic level, for example, consider the process of genetic recombination by which two DNA molecules exchange information, thereby generating new combinations of genes. Even in the accepted Darwinian model these small changes collectively are regarded as driving the evolution of the species over repeated generations. In other words, evolution is generally acknowledged to be the result of a large accumulation of small increments of order in an otherwise random process. Such ordering takes place at an unconscious biological level, perhaps even on the level of the collective unconscious of a species, and emerges from the uncertainty inherent in the complexity of all living systems. One can see this unconscious ordering reflected in the behavior of many communal species, i.e. swarms of insects, flocks of birds, schools of fish, etc., where the collective appears to have knowledge that is inaccessible to the individual. The observed variations in gender performance in the PEAR experiments, and the enhanced effects produced by resonant opposite-sex couples, may hint at a similar subliminal biological process.

An alternative evolutionary model was first proposed by Jean-Baptiste Lamarck some 200 years ago, and long rejected as heresy by most mainstream evolutionary theorists. He suggested that such ordering may not be purely random, but may incorporate a component of intention, albeit on an unconscious level, on the part of an individual or a species in response to environmental changes that is reflected in the transmission of acquired characteristics.[15] He claimed that organisms have a tendency to evolve toward increasing complexity through proactive responses to changing environments that require or encourage behavioral adaptation, not merely as the result of passive statistical events. Lamarck referred to this process as *le pouvoir de la vie* (the life force) or *la force qui tend sans cesse á composer l'organisation* (the force that perpetually tends to make order).

A similar view was put forth by the French philosopher Henri Bergson. His major work, "Creative Evolution" published in 1911,[16] posited a vital impulse he called *élan vital* that underlies the creation of all living things, a process of self-organization that he linked closely with consciousness. He maintained that there must be an impulse accounting for diversity and differentiation; that these tendencies can be defined as instinctive and intelligent; and that intuition (a function of the unconscious mind) allows us to place ourselves back in the original vital impulse.

Both the observation that organisms have a tendency to evolve toward increasing complexity and the concept of self-organization are inconsistent with the law of entropy, a dissipative process that drives closed physical systems toward increasing disorder. Less well known is its complementary principle of *syntropy*, which is postulated to produce a continuous increase in complexity, and draws individuals and systems together by their common characteristics and goals.[17]

Regardless of the model one invokes, the evolutionary process is acknowledged as being strongly dependent on the generation of randomness or "noise" emerging from the massive redundancy of continuously recombined genetic information. When the randomness of this process is constrained, as in repeated interbreeding, the short-term advantage of increased predictability of inherited traits can be offset by longer-term weakening of the genetic strain of the species.

Of necessity, all humans straddle the consciousness-imposed divide between the realms of subjective and objective, and thus embody a rich complexity that makes them capable of self-reference. In the complementary dynamic of self/not-self in their essential exchanges of information with their environments, they generate an inevitable degree of uncertainty, which, consistent with the implications of Kurt Gödel's Incompleteness Theorem, limits the precision with which they can represent themselves.[18] Survival of an individual or a species demands a degree of flexibility in the process of adaptation that is afforded by the intrinsic uncertainty in their interactions with randomness, and may even be enhanced by their exercise of some degree of subconscious volition to achieve an optimal outcome. Our consciousness-related anomalies may thus be nothing more, nor less, than indications of this fundamental ordering process.

Uncertainty and probability are intimately correlated, and the study of anomalies is clearly a probabilistic game. Indeed, uncertainty appears to be fundamental to any exchange of information between consciousness and its physical corpus, or with its ultimate Source, and may well be the raw material out of which any departures from chance are constructed.

The objective/subjective tradeoff itself imposes an inevitable uncertainty on any constructions of consciousness, ultimately constraining them to probabilistic expression.

The relationship between signal and noise also entails a degree of uncertainty that may be directly relevant to the manifestation of consciousness-related anomalies. Typically, the scientific method advocates systematic elimination of extraneous variables in order to bring the phenomenon in question into sharper focus. This was the assumption that originally guided the evolution of PEAR's remote perception program, but repeated refinements of the analytical techniques in order to refine the "signal" by reducing the "noise" had quite the opposite effect, reducing the signal to a progressively imperceptible level.[19] Similar departures from canonical expectations can be found in contemporary engineering applications known as "stochastic resonance," wherein a deliberate increase in the overall noise level in certain kinds of lasers or sensitive electronic circuits appears to enhance the detection of weak, fluctuating signals.[20] Other studies have demonstrated that the introduction of an element of chaos into various types of nonlinear processes, such as the interaction of two otherwise independent random oscillators, could stimulate synchronous behavior between the transmitter and receiver.[21] In either of these instances, information or order is being introduced into a sensitive nonlinear physical system, not by reducing the ambient noise, but by *increasing* it. It is worth noting that unanticipated corollary observations in these studies indicated that the receiver could record changes in the signal before the transmitter registered the transmission of those changes.

Indeed, it may well be that noise or uncertainty is the raw material out of which any anomalous effects are assembled. All of the human/machine studies at PEAR employed some form of random process and observed anomalous effects that appeared as small departures of their random outputs from chance, or small increments of increased order. If so, it would appear that the expression of such consciousness-related anomalies requires a certain degree of noise for the anomalous signal to manifest. And the demonstrated difficulties in the replication of reported effects also suggest that uncertainty may be a relevant component of their manifestation.

Uncertainty plays a major role in the world of quantum-level phenomena where most events tend to be complementary in nature, and thus allow two possible perspectives for making an observation, i.e., particle/ wave, position/momentum, time/energy, etc. Complementarity, a principle closely related to that of Uncertainty, holds that one cannot assign precise values simultaneously to complementary properties. In quantum mechanical formalisms, quantification of uncertainty is represented by the symbol h, also known as Planck's constant (6.626068 x 10^-34 m^2 kg/s (Joules/s). In the face of this elegant measurement and mathematical symbolism and its abstract application in quantifying "physical" processes, it is easy to lose sight of the fact that "uncertainty" actually refers to ambiguity, or a *lack of information*—a direct reflection of the state of mind of the observer, rather than a property of that being observed. It thus represents a form of noise, the same noise that appears to be a precondition for the appearance of anomalies.

There is another constant in nature, referred to as the Golden Ratio or the Divine Proportion, among other names, represented by the Greek letter φ, the earliest reference to which appeared in Euclid's *The Elements*.[22] Long recognized as an important geometrical ratio (1.618), it is obtained by dividing a line into two parts so that the longer part divided by the smaller part is also equal to the whole length divided by the longer part. This constant is related to the "Fibonacci sequence," where each number in a sequence is the sum of the two numbers that precede it, and has a broad range of applications in architecture, aesthetics, art, music, and spiral growth in nature. As nature's most ubiquitous fractal scaling ratio, this "spiral principle" can be observed in plants, animals, seashells, vortices of water and air, and many other natural phenomena, and it is present in the atomic to the galactic scales.[23] In 1854 Adolf Zeising, a German psychologist whose main interests were mathematics and philosophy, wrote of the golden ratio operating as a universal law

in which is contained the ground-principle of all formative striving for beauty and completeness in the realms of both nature and art, and which permeates, as a paramount spiritual ideal, all structures, forms and proportions, whether cosmic or individual, organic or inorganic, acoustic or optical; which finds its fullest realization, however, in the human form.[24]

He found this proportion expressed in the arrangement of branches along the stems of plants and of veins in leaves, in the skeletons of animals and the branching of their veins and nerves, in the proportions of chemical compounds and the geometry of crystals, and even in artistic endeavors. Here we have a somewhat more structured expression of uncertainty, one that appears to be associated with the organized development of biological as well as physical systems.

Yet another example of a biological ordering principle that might be considered is that of biophoton emission, a low-energy endogenous radiation produced by all living organisms and detected as barely visible light. Biophotons were first discovered by the Russian scientist Alexander Gurvich in 1923, but it wasn't until the 1970's that German biophysicist Fritz-Albert Popp and his colleagues carried out extensive studies that indicated that biophoton light is stored in the DNA of all organisms.[25] The degree of coherence of this optical field has been shown to be an accurate indicator of the health of a living system, *e.g.* emissions of diseased or dying organisms tend to be more chaotic. Popp proposed that the coherent biophoton field may represent a form of intracellular communication that serves to regulate life processes such as morphogenesis, growth, differentiation, and regeneration.

It is not clear just how these various concepts relate to the manifestation of consciousness-related anomalies, but they do provide supporting evidence for the existence of an innate biological ordering principle that utilizes the random noise in living systems. As we observed in our book *Consciousness and the Source of Reality*,[26] the empirical results of the PEAR studies have led us to believe that these phenomena are deeply rooted in, and therefore significantly indicative of, a more fundamental, profound, and ubiquitous metaphysical dynamic whose ultimate comprehension holds far richer potential for human benefit than mere explicit phenomenological curiosities. In brief, as our culture has become increasingly captivated by a materialistic world view, it has gradually lost sight of its connection with the more transcendent dimensions of reality.

Many traditions speak to this connection between the physical world and the Source. It is recognized in many cultures and known by many names, *e.g.* Chi or Qi, Archeus, Vital Force, or Prana, among others. Yoga, for example, is an ancient practice that is believed to have its origins in pre-Vedic Indian traditions. The Sanskrit word yoga roughly translates as "to add, join, unite, or connect." While popularly regarded as a form of physical discipline involving various stretching exercises or poses, the primary goal of yoga actually is to integrate body and mind to achieve unity with the Source through the flow of prana.

A more contemporary model of this connection is offered in G. Spencer- Brown's *Laws of Form*,[27] wherein he maintains that all emergent phenomena stem from the act of distinction from the primordial Source, or what is referred to in some traditions as the "Void." This concept is consistent with the premises of virtually all religious, mystical, and philosophical traditions, which maintain that the act of separation from the Divine lies at the heart of all creation. Such separation or distinction, however, is an act of consciousness and does not negate the underlying unity between the Source and Its creations since, as the ancient alchemists were fond of observing, "That which is below is like that which is above, and what is above is like that which is below, to accomplish the miracles of one thing."[28]

The phenomena we term "anomalies" are mysterious primarily because they cannot be explained by any known physical model, but this well may be because they are not inherently physical in nature. Rather, they represent a domain where the objective physical world and the subjective experiential world overlap, and they hint at a connection to the primal Source from which all phenomena emanate. Our thesis is that the body, through the agency of the life force, may represent that intersection through a more direct connection with the Source than the mind, or at least the conscious mind, and may be the ground from which intuition and creativity arise. This dynamic reflects the Taoist principle of the balance, or complementarity, of the fundamental principles of yin and yang, the receptive and the active forces of the universe.

While we concede that this proposition is speculative, nonetheless it is consistent with many non-reductionist traditions and is supported by the empirical evidence of carefully controlled laboratory studies. At the very least, it is clear that modern science needs to be expanded to accommodate intangible subjective experience, and that the prevailing notion of "consciousness" must move beyond the limited scope of the physical brain to include the "underworld" of the unconscious and the life force itself.

Endnotes

1 Bergson, H. (1911). *Creative Evolution*. Tr., Arthur Mitchell, New York: Dover, 1998.

1 Jahn, R.G. and Dunne, B.J. (2009). *Margins of Reality: The Role of Consciousness in the Physical World*. Princeton, NJ: ICRL Press, p.203. (Originally published by Harcourt Brace Jovanovich in 1987.)

2 Jahn, R.G. and Dunne, B.J. (2011). *Consciousness and the Source of Realty: The PEAR Odyssey*. Princeton, NJ: ICRL Press, 2011.

3 Jahn, R.G. and B.J. Dunne (2005). "The PEAR Proposition." *Journal of Scientific Exploration*, 19 (2) pp. 195–246.

4 Jahn, R.G., Dunne,,B.J, Nelson, R. D., Dobyns, Y. H. and Bradish, G. J. (1997)." Correlations of Random Binary Sequences with Pre-Stated Operator Intention: A Review of a 12-Year Program." *J. Scientific Exploration*, 11, No.3, pp.345–367.

5 Dunne, B.J. and Jahn, R.G. (2003). "Information and Uncertainty in Remote Perception Research." *Journal of Scientific Exploration*, 17 (2) pp. 207–241.

6 Radin, D. (2013). *Supernormal: Science, Yoga, and the Evidence for Extraordinary Psychic Abilities*. New York: Deepak Chopra Books (an imprint of the Crown Publishing Group of Random House, Inc.).

7 Peoc'h, R. (1995). "Psychokinetic action of young chicks on the path of an illuminated source." *J. Scientific Exploration*, 9: 223–229.

8 Bird, C. (1997). *The Divining Hand: The 500 Year-old Mystery of Dowsing*. Atglen, PA: Whitford Press.

9 Jahn, R.G. and Dunne, B.J. (2001). "A Modular Model of Mind/Matter Manifestations (M5)." *J. Scientific Exploration*, 15: 299–329.

10 Lokhorst, G-J. (2008). "Descartes and the Pineal Gland". In Edward N. Zalta, Ed. *The Stanford Encyclopedia of Philosophy* (Summer 2011 Edition). Lokhorst quotes Descartes in his *Treatise of Man*.

11 Descartes, R. (1985). In J. Cottingham, R. Stoothoff, and D. Murdoch, Trans. *The Philosophical Writings of Descartes, Volume 1*. Cambridge University Press, p.341.

12 Huxley, A. (1954). *The Doors of Perception*. UK: Chatto & Windus.

13 Jahn, R.G. and Dunne, B.J. (2009). *Margins of Reality: The Role of Consciousness in the Physical World*. Princeton, NJ: ICRL Press, p.203.

14 James, W. (1011). *Some Problems of Philosophy*. NY: Longmans, Green, & Co., p.50.

15 Lamarck, J. B. (1914). *Zoological Philosophy: An Exposition With Regard to the Natural History of Animals*. London: Macmillan & Co.

16 Bergson, H., (1911). *Creative Evolution*. Tr., Arthur Mitchell, New York: Dover, 1998.

17 DiCorpo, U. and Vannini, A. (2014). *Syntropy: The Spirit of Love*. Princeton, NJ: The ICRL Press.

18 Gödel, K. (1931). "Über formal unentscheidbare Sätze der Principia Mathematica und verwandter Systeme, I", *Monatshefte für Mathematik und Physik*, v. 38 n. 1, pp. 173–198.

19 Dunne, B.J. and Jahn,R.G. (2003). "Information and Uncertainty in Remote Perception Research." *Journal of Scientific Exploration*, 17 (2) pp. 207–241.

20 McNamara, B. and Wiesenfeld (1988). "Theory of stochastic resonance." *Physical Review* A, 39, No 9, pp.4854–4869.

21 Jones, R.J., Rees, P., Spencer, P.S. and Shore, K.A. (2001). "Chaos and Syncronisation of Self-Pulsating Laser Diodes." *Journal of the Optical Society of America*, B., No.2, pp. 166–172

22 Euclid, (300 BC). *The Elements*. In Ball, W.W. Rouse (1960). *A Short Account of the History of Mathematics* (4th ed. [Reprint. Original publication: London: Macmillan & Co., 1908] ed.). New York: Dover Publications.

23 Cook, T.A. (1914). *The Curves of Life: An Account of Spiral Formations and Their Application to Growth in Nature, To Science and to Art.* London: Constable and Co.

24 Zeising, A. (1854). *Neue Lehre van den Proportionen des meschlischen Körpers.* Preface.

25 Popp, F. A., Quao, G., and Ke-Hsuen, L. (1994). "Biophoton emission: experimental background and theoretical approaches." *Modern physics Letters B*, 8 (21–22).

26 Jahn, R.G. and Dunne, B.J. (2011). *Consciousness and the Source of Realty: The PEAR Odyssey.* Princeton, NJ: ICRL Press.

27 Spencer-Brown, G. (1969). *Laws of Form.* London: Allen & Unwin.

28 Hermes Trismegistus, *Emerald Tablet,* in J. Ruska, *Tabula Smaragdina* (1926). Heidelberg: Carl Winter's Universitätsbuchhandlung, p.2.

Orgone Energy and the Life Force

RICHARD A. BLASBAND

I am well aware that the human race has known about the existence of a universal energy related to life for many ages. However, the basic task of natural science consisted in making this energy usable. This is the sole difference between my work and all preceding knowledge.
— WILHELM REICH[1]

The concept of a life energy or life force that animates life has been with us since antiquity in the Chinese concept of "chi," and the East Indian concept of "prana." In pre-modern times Aristotle thought that it was necessary in understanding the origin of life. He asked, "Why cannot the seed at its origin be so created, that it can turn into blood and flesh without itself having to be blood and flesh?" and "How do these parts come into being: does the one form the other or do they simply arise one after the other?" His answer was the soul. "The soul cannot exist without a body, yet it is not the body but rather something inherent in the body." It is "the principle of all things living... If the eyes were a living being, then eyesight would be its soul, this being the substance as notion or form of the eye; and the eye would be the matter of the eyesight." Aristotle maintained that the mere stuff or matter is not yet the real thing; it needs a certain form or essence or function to complete it. Matter and form, however, are never separated; they can only be distinguished. Thus, in the case of a living organism, for example, the sheer matter of the organism (viewed only as a synthesis of inorganic substances) can be distinguished from a certain form or function or inner activity, without which it would not be a living organism at all; and this "soul" or "vital function" is what Aristotle in his *De anima* (On the Soul) called the entelechy (or vital principle) of the living organism. Aristotle's assertions held sway until well into the 17th and 18th centuries.

In the pre-modern era, prominent physicians and scientists such as Harvey, Stahl, Wolff, and Blumenbach wrestled with the concepts and tried to find solutions to the prevailing problems of the origin of life, conception, evolution, and biological development. A variety of terms were created to explain these events, including "phlogiston" and "vis essentialis" and "vita propria," which Blumenbach maintained established the specific vital activity of the systemic parts of the living organism. According to Hans Dreisch, who wrote a scholarly book on the subject, Blumenbach's contribution was the only one of the pre-moderns who made a significant step in our understanding of the subject since Aristotle. Indeed, it was the first to establish what we consider as modern physiology.

From a purely philosophical point of view, Kant, Schopenhauer, Bergson, Schelling, and Hegel had much to say on the subject of vitalism, as it was so intrinsically part of the structure of human thought as an explanatory principle in understanding life. Their thought was bound up with notions of "energy," "autonomy of life," teleology, intention, determinism, and the "soul," and inevitably, their influence on what became mechanistic principles of development and movement. It was difficult to wrestle with the concepts and their interrelationship. Indeed, according to Dreisch, many critiques of the concept of vitalism, such as those of Bernard, when examined in detail reveal themselves to be essentially vitalistic in nature.[2]

The advent of Galileo's (mid-1500s to mid-1600s) work wrought major changes in how we conceive of ourselves and our universe. He championed a quantitative, analytic view of natural processes which dominates most scientific thought to the present time. In essence, he and those devoted to mathematical physics found nature to be a process whereby exclusively mechanical principles were at work. This included, of course, the theory of life. The appeal of mechanism was that much could be done with it both theoretically and practically. Little could be done with so-called "vitalistic" theories, and they began to wither on the vine.

Not all, however, succumbed to mechanism as a final explanatory principle.

The chemist, Justin Liebig, (1803–1873), unable to satisfy his understanding of life through strictly mechanistic principles, was a prominent vitalist, maintaining that a "fourth cause" is found only in living beings and dominates the forces of cohesion and combines the elements in new forms so that they gain new qualities. And as we come deeper into the modern era, we find the work of the French physiologist, Claude Bernard, who died in 1878, who maintained that the growth of individual forms was a function of the "whole" whose laws are initially dormant and then activated with growth and development.

Darwinian evolution to the present day is dependent upon mechanistic thinking. Jean-Baptiste Lamarck (1744–1829) had offered an alternative, more functional view, but it never gained traction in science until just recently, where we find an interest in his concepts based on experimentation.

The decline of vitalism during the period just described had an interesting parallel in the development of laws of human behavior. Psychology and biology, being so intertwined, the doctrine of psychophysical parallelism flourished contemporaneously with materialistic science. Human action was thus also subordinated to the general materialism; any natural event was a mechanical event. A "soul" or its equivalent is not considered to be an element in the causality of nature.

Underground, vitalism survived through to the present time despite the overwhelming publication of papers and books in the scientific media based upon mechanistic principles. The names of the vitalists are many and their efforts to survive valiant. They began to enter the experimental field with the demonstration of alleged life energies by Mesmer (vital fluid) and von Reichenbach (Od), but it was not until the work of Wilhelm Reich that experimental evidence of a life force manifesting in so many varying domains of nature that one felt that a true and deep breakthrough had taken place.

Reich became aware of the existence of this energy, which he called "orgone," not through cosmology or astronomy, but through his studies in sexuality, specifically the function of the orgasm. As a psychoanalyst practicing in Vienna in the 1920s, he found that the elimination of neurotic symptoms followed rapidly upon the patient's capacity to experience a satisfying discharge of all sexual tension during the genital embrace. Following Freud's initial hypothesis, Reich realized that there was an economy to the libidinal charge within the organism. Through food stuffs and exposure to the atmosphere we build up an energetic charge, which must be discharged at periodic intervals lest potential neurotic complexes become charged and neurotic behavior ensues. Discharge could be effected by any of several sexual practices, work, menstruation, and childbirth, but was most satisfactory on a regular basis in the genital embrace with a loved partner.

For Reich, the discovery of the "orgastic plasma pulsation" was, like Columbus's discovery of America, "...the coastal stretch from which all else developed."[3]

Bioelectricity

"Something" was discharged and Reich set out to understand what it was. He succeeded in objectifying the phenomenon by measuring bioelectrical charge on the skin surface when subjects were in emotional states: sadness, anxiety, and sexual pleasure. The oscillograph consistently demonstrated that states of anxiety were always associated with a drop in charge on the skin surface, states of pleasure with an increase of charge, and anger with a blocking of charge at the muscular apparatus.[4]

The bioelectrical experiments verified something that Reich had seen when working purely clinically: The organism is in a constant state of pulsation, of expansion and contraction, of reaching out to its environment in a state of biophysical expansion or contracting away from it, much like an ameba extending a pseudopod towards food, a comparison that Freud had made some years before.

Reich had hoped that one thing that the bioelectric experiments would reveal is the nature of the life energy. Extant at the time was a bioelectrical theory of life, which Reich had embraced. With the bioelectric experiments he realized, however, that electrical energy as it was known at that time had an irritating effect on living substances, therefore it could not be the life energy per se. He began a series of experiments searching for the source of life energy by examining boiled food stuffs under a microscope, reasoning that food and some form of energy from the atmosphere were the primary sources of energy for living processes. He also examined single celled organisms to understand the processes of pulsation that he had seen in patients and in the bioelectric experiments.

Bions

When Reich examined boiled foodstuffs he found, to his surprise, that with extensive boiling no matter the food source, fats, proteins, carbohydrates, they all broke down into similar forms. These were microscopic vesicles that moved from place to place, pulsated, and had a bluish glow. The vesicles, which he named "bions," were two to three times the size of bacteria, but could not be bacteria because of the extensive boiling process. And, unlike bacteria, which move rapidly across the field of vision under the microscope, the bions moved only a short distance from place to place about a central point.[5]

Reich found that he could culture the bions on standard media under strictly sterile conditions. Bion cultures strongly radiated something which could be felt as a tingling sensation on the skin or registered as a burn of the skin when directly applied. The laboratory space began to glow with a bluish radiation, film packets registered anomalous light impressions and metallic objects in the laboratory became spontaneously magnetized. Tests for radioactivity at a local hospital proved negative. Following a series of experiments testing the effect of bion preparations on electroscopes and other devices that detect "static electricity," Reich realized that he had discovered an energy or force that originated in living substance, was released from living substance, and could exert an impression on a variety of substances. To his mind this was the life energy he was seeking. Further research fortified this impression: tests of the energy's healing qualities were positive.[6] Reich was also able to create bions by heating inorganic materials such as carbon, iron, and ocean sand (silicon) to incandescence and culturing the heated, sterile product in sterile media. The result in the case of sand was large packets of bions with a strong glow and the capacity to burn the skin through a glass test tube.

Reich's basement laboratory in Oslo in mid-Winter exhibited anomalous light phenomena—bluish fogs and small lightening-like streaks of light. Thinking that these were due to the presence of the numerous petri dishes filled with sand bion cultures Reich placed cultures into a box made of alternating layers of non-metallic and metallic materials with the metal on the inside. He hoped that the non-metallic material would prevent the radiation, whatever it was, from leaving the box, while the metal would reflect the radiation back in toward the center of the box. A glass wall permitted observation of the radiation. Reich did, indeed, see the radiation in higher concentration, but was puzzled when it persisted even after removing the cultures from the box. He then realized that the radiation was everywhere, but concentrated in the box by virtue of the materials in its structure. This radiation was named "orgone energy" from the fact that its discovery came about from Reich's original studies of the function of the orgasm and because the radiation could be absorbed by organic materials. The enclosure was named the orgone energy accumulator. Orgone energy was life energy as manifested in the atmosphere. Later experimental work and theoretical deductions convinced Reich that it was also the cosmic energy, per se.[7]

Reich devoted the remaining years of his life to experimental work with physical orgone energy. He discovered that it had an anti-entropic quality, was antithetical to electromagnetism, could reduce tumors and increase longevity in cancer-affected mice and ameliorate cancer in humans,[8,9] run motors, nullify and transform radioactivity, create and dissolve clouds through manipulation with long metal tubes, and be transformed into variants that had noxious qualities. These discoveries were published in his journals, the Orgone Energy Bulletins, and many of the experiments have been replicated by responsible scientists worldwide.[10]

Consciousness

One of the major problems facing those who study consciousness is the Cartesian split between the mind and the body ever since Descartes had described it in 1644 in a famous philosophical treatise. A declaration of independence, it freed rational science from the mystical influence of the Church. In doing so, however, scientists and philosophers saw man as having a "split" between mind and body. Writing about this, in 1964 the eminent biologist Ludwig von Bertalanffy said, "Do not say that the Cartesian dualism is a dead horse or a straw man erected to be knocked down, as nowadays we have 'unitary concepts' and conceive of man as a 'psychophysical whole.' These are nice ways of speaking, but as a matter of fact, the Cartesian dualism is still with us and is at the basis of our thinking in neurophysiology, psychology, psychiatry, and related fields."[11] This hasn't changed.

Reich made major contributions to resolving the Cartesian split by his clinical studies and experimentation. Clinically and practically Reich provided an entry into the direct treatment of neuroses and psychosomatic illnesses by manually working with the somatic muscular armoring, the complementary aspect of psychic character armoring. His bioelectric studies provided an experimental proof of the relationship between subjective qualitative experience (the mind) and objective quantities (the body). He found, for example, that the stimulation of an erogenous zone, such as a woman's nipple, would provide a perception of pleasure *only if there was an increase in bioelectrical potential at the nipple.* Stimulation without this effect, even if the nipple was erect, would not result in the perception of pleasure. The quality of the stimulation was important as was the readiness of the person to be aroused. As Reich stated, "*In order to make the pleasure sensation perceptible, there is necessary, in addition to the mechanical congestion of the organ, an increase in bioelectrical charge. The psychic intensity* of the pleasure sensation corresponds to the *physiological quantity* of the bio-electrical potential" (italics Reich's).[12]

This was a great advance in understanding the nature of the psycho-somatic relationship. While many corrélations between the two had been generally postulated in mainstream bio-medical research, this was the first time, to my knowledge, that such an immediate *complementary* relationship had been experimentally demonstrated.

Reich saw the *sensation* of pleasure to be the functional pair of the *quantity* of bio-electric charge. The common functioning principle of the antithesis between these two paired functions was never described by Reich, except in the general sense that the two were functionally identical with respect to streaming orgone energy in the organism. In a lecture at New York University I offered the idea that the two were to be found as one in the energy *field* of the organism, having in mind at that time the work of Harold Burr, professor of anatomy at Yale University who, utilizing a DC field meter of his own design in the 1930s. found such fields governing manifold aspects of biological functioning in a variety of plants and animals. Although he was measuring direct currents, Burr did not believe that the "force" at work in plants and animals was electricity per se. Hence he defined it as an "electrodynamic" quality, without describing further what he meant by the term. He was, however, able to demonstrate that the bioelectric field of an amphibian egg had a direct organizing effect on the establishment of the geometry of the spinal cord of its generated fetus.[13]

Reich never made deeper inroads into describing consciousness. It is my view that this was, in part, due to his views on spirituality derived from his experiences as a physician in Vienna in the 1920s. There he saw many people attending spiritual séances, which he thought was a "running away" from their facing reality. When he threw out spiritualism he dumped the baby, the "spirit,"[14] with the bathwater.

In his discovery of orgone energy, however, Reich did make singular contributions to our understanding of "life energy," which he found to be anti-entropic (syntropic[15]), excitable both spontaneously and in response to external stimuli, luminous, healing, a motor force, the source of all life functions, and as a powerful operative force. Above all, he established a foundation for experimental investigation of the most intriguing questions affecting nearly all scientists no matter what their fundamental driving interest.

Endnotes

1 Reich, Wilhelm in Kevin Hinchey's film, *Man's Right to Know*, Wilhelm Reich Infant Trust, 2002.
2 Dreisch, H. *The History and Theory of Vitalism*, (C. K. Ogden, trans.) London: Macmillan, 1905.
3 Reich, W. Ether, *God, and Devil*. Orgone Institute Press, Rangeley, 1949.
4 Reich, W. *The Function of the Orgasm*. Orgone Institute Press, N.Y., 1942.
5 Reich, W. *The Cancer Biopathy*. NY: Orgone Institute Press, 1948.
6 Reich, W. ibid
7 Reich, W. ibid
8 Reich, W. ibid
9 Blasband, R. "The Orgone Accumulator in the Treatment of Cancer in Mice," *Subtle Energies and Energy Medicine* 20:2, 2009.
10 For a comprehensive and detailed review of Reich's and others' biophysical and weather work see Maglione, Roberto, *Methods and Procedure in Biophysical Orgonometry Imiolibro*. it, 2012 and Wilhelm Reich and *Healing of Atmospheres*, Natural Energy Works, Ashland, 2007.
11 Von Bertalanffy, L. "The Mind-Body Problem: A New View," *Psychosomatic Medicine* 24:1, 1964, p. 29.
12 Reich, W. *The Function of the Orgasm*, p. 329.
13 Burr, Harold, S. "The electro-dynamic theory of life," *The Quarterly Review of Biology*, 10:3, 1935.
14 For a cogent and scientific explanation of spirit see the writings of the master healer, clairvoyant, and theoretical physicist, Nicolai Levashov (Levashov 1994, 1997, 2000).
15 Di Corpo, U. and Vanini, A., *Syntropy: The Spirit of Love*. Princeton NJ: The ICRL Press, 2015.

The Nature of Mind and its Phylogenetic Origins*

ANTONIO GIUDITTA

Introduction

According to present scientific knowledge, mental activity is assumed to be fully reducible to brain activity that takes place at the molecular, cellular and system levels. This reductionist view is in contrast with the monistic concept that considers mental and physical features to be different aspects of the same substance. Reliable hypotheses on the nature of the human mind should account for all its major features/capacities, including the strict association with the brain, generation of subjective images, learning and memory processing, and involvement in paranormal phenomena. Accordingly, by logical inferences from these features and indirect but objective criteria, we have examined the presence of mind in phylogenetic primitive species and inanimate matter, and have obtained evidence that mental features are present in all organisms, including archeobacteria and prokaryotes, and that they may be traced back to elementary particles.

The mind/body problem has generally been examined by limiting considerations to the mental faculties of adult man. This approach has marginalized the problem of the phylogenetic origin of the human mind, by burying it under inquiries into the origin of the human brain, as prescribed by the accepted correlation between mental activity and brain activity. This reductionist attitude is, at best, misleading. Hence, in view of the relevance of the subject, we have undertaken an independent evaluation of the origin of the human mind by approaching an old problem from a novel point of view (Giuditta, 2004).

The biological features of organisms are well known to emerge and progressively differentiate in the initial period of life cycles (ontogenesis),

* This article was originally published in *Human Evolution 27: 281-290, 2012.*

at a time when specific traits are established according to genomic instructions and epigenetic modulation. These primary stores of biological information have been dynamically acquired and systematically laid down during the course of biological evolution lasting 4-5 billion years and following a much longer prebiotic period. This basic body of knowledge implies that a satisfactory understanding of the genesis of mental and material features of organisms may only be attained by investigating progressively unfolding phylogenetic events.

There is a multitude of names encompassing the concept of mind: soul, spirit, thought etc., each with its own specific connotations, that may be veiled by its tacit inclusion in the concept of mind. An additional qualitative distinction is needed between the mind of man and the mind of God. To discuss the mind of God may be a productive exercise, but only to point out some of the contradictory features that the human mind insists on assigning to it. By contrast, we do have first-hand information on the features of the human mind that allow more direct considerations.

Let us start with the evident, but at times neglected, distinction between the conscious and non-conscious mind. Clearly, non-conscious mind extends over vital domains that are much wider and deeper than those of the conscious mind. They actually allow the conscious mind to exist and operate. Boundaries of the non-conscious mind extend to the very source of life.

People have argued about the features of mind for centuries. Theologians and philosophers were the first to undertake this endeavor, then came science experts. Among a wealth of virtues, flaws and trivial considerations, most claims from this virtual and protracted roundtable have concerned thought processes and states of mind from the perspective of the onlooker and the acting subject. I am unaware of specific proposals for the investigation of the nature of mind, with the exception of those made by scientists with a reductionist connotation, and by theologians and philosophers who claim that the mind is unknowable since it is a substance ontogenetically distinct from the physical world.

Personally, I believe that any speculation on the nature of mind should be postponed until a comprehensive list of its features/capacities has been prepared to the best of our knowledge. What should be avoided are proposals that only concern specific features of the mind, a temptation which is often encountered when treating systems or processes of great complexity that may only be conceived in their entirety with the utmost difficulty. With regard to the concept of mind, it may not be possible to attain an exhaustive view because our quest is an introspective query into our own mind. Nevertheless, the benefit of undertaking such an exercise cannot be excluded *a priori* even if the observer and the observed share similar dimensions and the same nature.

Features/capacities of mind

The following list of features/capacities of the human mind is, most likely, approximate and incomplete, and is not presented in a priority order (Giuditta, 2004; 2011).

(a) The human mind is an integral part of the human body, and is specifically related/identified with the human brain. Body and brain are physical correlates of mind endowed with a highly complex structural organization. They are well defined biological objects that have been phylogenetically determined. As a result, while the association of mind with body/brain may not necessarily exclude the existence of disembodied minds, the notion raises the issue of the philogenetic origin of the human mind. It is not just a question of journeying backwards in time to the very beginning of the evolutionary history of living organisms (a convenient albeit arbitrary point of departure), but of proceeding even further back, to reach and consider the condensed, adimensional situation assumed to exist just before the big bang.

The strict association of mind with brain implies that the phylogenetic origin of the human mind may be strictly related to the phylogenetic origin of the human brain. It thus follows that all organisms with a brain, including those exhibiting progressively less evolved and more primitive brains or nervous systems, should be considered to display mental features. The latter capacities are to be considered correspondingly less complex and more primitive than those of man and higher animals, and may be presumed to be partially or largely subconscious. This line of reasoning extends the concept of mind backwards to all phylogenetic lineages characterized by a nervous system, down to the limit of single nerve cells. This position is at variance with views envisaging the presence of mind only in higher animals. To the contrary, it greatly extends mental aspects to phylogenetic periods not previously considered to harbor mental features.

(b) Mind is capable of representing the structure of the outside world with the qualities, interrelations and complexities pertaining to our immediate environment, but often including remote space and time domains (from infinitesimally small regions to astronomic dimensions). Perceptions arise from specialized sensory receptors which are selectively sensitive to mechanical, chemical, thermal and electromagnetic forms of energy that originate in the external world and within the body. Inputs from these receptors reach the central nervous system and combine to generate the numberless qualities of mental images (*qualia*) we communicate to other human minds, albeit in an unrefined way. Representations give birth to concepts and logical connections, and lead to the emergence of rational thought. In so doing, they leave modifiable traces (memories) contributing to the quality of our mental states, including dreams and nightmares.

A curious enigma is how the transduction of various forms of stimuli encoded into comparable patterns of neural impulses and chemical signaling succeed in generating mental images of extraordinary diversity. As these images are clearly to be listed among the most basic mental events, they may hardly be reduced to brain activity notwithstanding their strict association with it. Take, for instance, electromagnetic waves impinging on the retina and generating forms and colors. They activate rhodopsin and visual circuits by physical mechanisms in no way differing in nature from incoming physical stimuli. As such, they remain markedly different from *qualia*.

Activated visual circuits transfer information across brain regions, notably from receptors to the neocortex. The operating bioelectric mechanism remains the same irrespective of the type of activated receptor, be it acoustic, visual, chemical, or mechanical. Such consistent uniformity makes it unlikely that the high specificity of *qualia* may solely and confidently be attributed to the local distribution of neural activity. On the other hand, whatever further condition may be required for the generation of *qualia*, it should not be equated to the notion of immaterial 'spirit' postulated by religions and philosophical systems. Due to the strict temporal coincidence of mental images with brain events, such a putative solution requires the additional *ad hoc* assumption of two ontogenetically distinct events (*qualia* and brain activity) coinciding in time, while fully independent of each other. The assumption lacks logical support and appears unwarranted.

Whatever mechanism may eventually be shown to be involved in the generation of *qualia*, their strict association with neural activity raises the legitimate question of the phylogenetic origin of such a remarkable link. Was it active since the first appearance of a nervous system, or has it only emerged at a certain level of neural complexity? In the latter case, how could one identify the level of complexity below which mental events would be definitely absent?

Before embarking on the quest for plausible answers, let us first consider the following thoughts. The existence of *qualia* bears on the basic distinction of 'self' from 'non-self'. Only entities included in the 'self' category are assumed to experience *qualia*, since *qualia* appear to only belong to a subjective 'self' (so to speak, they are limited to his inside). Indeed, any observer (a 'self' in his own right) cannot perceive *qualia* perceived by a different 'self'. He remains on the outside of that 'self' and belongs to his 'non-self'. Yet, despite this substantial difficulty, each human observer knows that every other man is experiencing *qualia*. Somehow, this capacity is implied by the presence in all individuals of sensory receptors and appropriate responses. It follows that if such features were to be identified in any other organism, they should be considered a reliable sign of the capacity to generate mental images.

Indeed, this inference may be used as an objective criterion in the search of mental capacities in non-human organisms, regardless of lineage and physical complexity. Results of this phylogenetic survey appear clear-cut and unexpected, as all organisms exhibit minimal biological features consistent with the capacity to generate *qualia*. Indeed, all organisms (archeobacteria and prokaryotes included) are endowed with sensory receptors and exhibit physiological/behavioral responses congruous with their stimulation. Their *qualia* should clearly be considered much less refined and much more primitive than those of man, primates, and higher mammals and birds, but still appropriate to the different environments and life styles of different species. Even such primitive *qualia* are to be listed in the same category of mental images perceived by man and higher animals.

Are *qualia* of primitive organisms conscious, or are they largely or entirely subconscious? This question is likely to remain without a satisfactory answer. Even the existence of subconscious *qualia* may be questionable. Yet, the dynamics of our subconscious drives confirm that subconscious representations exist, undergo processing, and influence our conscious experiences. The obvious truth is that *qualia* of all organisms (man included, let alone primitive organisms) are not consciously available to external observers, but this may not be a relevant difficulty. Each human being is unaware of *qualia* experienced by other human beings. Yet, all of us know that other people have minds.

Since archeobacteria and prokaryotes are the simplest and earliest forms of life, the generation of basic mental events such as *qualia* should be considered a capacity shared by all organisms. Clearly, their degree of sophistication and refinement is likely to be very low with respect to human standards, and should be thought to be proportional to the physiological/behavioral complexity of the species in question. It follows that the mental capacities attributed to species endowed with a nervous system should be extended and shifted further back by a long stretch to reach a phylogenetic period preceding the emergence of eukaryotic cells.

(c) The conscious mind is capable of well-pondered choices and takes command of bodily movements by the special type of potential energy we call our will. In comparison with these extraordinary capacities, the largely unfathomable faculties of the non-conscious mind stand out even more impressively. They allow the integration and coordination of external and internal body movements and activities that evolved through phylogenesis and were acquired during the life cycle. Indeed, the mind is capable not only of processing incoming information from environmental sources and the organism's responses, but also successfully solving problems arising from the organism's interactions with the environment. Problem solving is comparable to the generation of a meaningful mosaic of informational tesserae, partly explicit, partly retrievable by data processing. Learning and problem solving may be considered cybernetic processes from the viewpoint of mechanisms, but may also be traced back to mental operations when subjectively perceived (Giuditta, 2008).

The capacity to learn is present in all species, even in species lacking an overt nervous system, such as prokaryotes, archeobacteria, and plants. Bacteria may learn to express genes promoting aerobic or anaerobic metabolism depending on the combination of environmental variables to which they are exposed (Tagkopoulos et al., 2008). Complex modifications of the structure of DNA that allow its transfer from the micronucleus to the macronucleus of ciliate archeobacteria are properly described by algorithms (Ehrenfeucht et al., 2007). Primitive unicellular and multicellular species, and plants, are capable of acquiring and processing information (Baluška & Mancuso, 2009; Darwin, 2010). On the whole, computing operations that lead to learning and problem solving appear to be occurring in all organisms. While their subjective dimensions cannot be verified, the analogy with human processing operations suggests that primitive mentation related to learning and problem solving has been present since the beginning of biological evolution, presumably at a subconscious level.

To speak of subconscious mentation does not imply a contradiction in terms. Our conscious activity consistently requires a great deal of subconscious operations, only a fraction of which may become conscious. Subconscious operations support learning as well as conscious operations. In fact, i) sleep processing is absolutely required for memorizing complex tasks (Ambrosini & Giuditta, 2001); and ii) problem solving is more efficiently attained during sleep than while awake (Wagner et al., 2004). It may be safely concluded that we are usually not aware of subconscious activity, but if we are asked to decide whether that activity is conscious or subconscious *per se* (that is with regard to the system harboring it), the question remains unanswered and generally ignored.

(d) In previous paragraphs we have left unanswered the question of whether mental features may be discerned in inanimate objects. If the problem is viewed from the perspective of Cartesian duality, one should unavoidably conclude that the inanimate world only contains mindless matter, whereas mind and matter happily coexist in the biological world. Conversely, if the problem is viewed from the perspective of a unitary world, additional considerations should be examined and brought forward to extend the mind domain below the biological threshold.

If the former duality is accepted, the mind of living organisms is bound to be conceived as an emergent feature, i.e. the appearance of novel capacities in entities generated by subunits lacking them. The simplest pertinent example is provided by water, whose properties are substantially different from those of its constituent hydrogen and oxygen atoms. It should be pointed out, however, that molecular properties reflect the spatiotemporal distribution of the component electrons and nuclei, and that this distribution is radically different in water from that of hydrogen and oxygen atoms. Hence, novel qualities emerging in water and other molecules only testify to the prodigious creativity of subunits that are able to produce countless different configurations by dynamically rearranging themselves. The same reasoning should be extended to the molecular subunits whose properties are likewise accounted for by the combinatorial rearrangement of their underlying subunit level. Hence, we are progressively pushed backwards down to the elementary particles that emerged after the big bang and appear to be the ultimate source of phylogenetic creativity.

Could immaterial mental events be attributed to elementary particles? In principle, and perhaps surprisingly, the answer appears to be positive, since elementary particles share properties (energy fields, uncertain wave-particle identity) that have little in common with inanimate objects identified by classical physics (definite identity, defined boundaries). Indeed, they rather resemble the properties of immaterial entities. It follows that, given the basic role of elementary particles in the generation of inanimate (and animate) bodies, it is not unreasonable to assume that mental features may also occur in all inanimate objects as well.

This remark has taken us, almost unexpectedly, into the core of the unitary world perspective. We are bound to proceed along this course in the attempt to adapt previous propositions to complexity levels well below the biological threshold. Above it, interactions of stimuli with organisms have been considered crucial steps for the generation of *qualia*. On the other hand, below the biological threshold such interactions need to be stripped to their bare essentials to make them congruent with the features of inanimate bodies.

It is impressive that one of the most relevant features of inanimate bodies is their never ending mutual collisions and interactions, often producing relevant modifications. The nature of their interactions is in no way different from that of interactions occurring between stimuli and sensory receptors. Hence, the interactions of inanimate bodies may also be thought to generate mental images, albeit of a substantially more primitive quality. It cannot be forgotten that in the evolving universe any modification of existing entities (from elementary particles to people) requires the interaction with suitable partners. This is particularly true when the modifications yielded increments in complexity. From this perspective, man's subjective experience may well turn out to belong to the same category of experiences made by colliding inanimate particles. There seems to be no sound reason to consider them purely material objects.

Accordingly, 'self' and 'non-self' qualifications respectively applied to organisms and impinging stimuli become more blurred when regarding inanimate objects. In their case, by applying the terminology applied to a living organism, each interacting partner should be considered a 'self' that is stimulated by a partner, but also a 'non-self' that stimulates its partner. It follows that 'non-self' may remain a superficially accepted qualification of stimuli interacting with organisms, but an inappropriate qualification when regarding the mutual interactions of inanimate bodies concurrently playing the 'self' and 'non-self' roles.

(e) The fifth consideration stems from the postulated unity of the universe with its mental and physical features. Since elementary particles are considered to be the first entities emerging from an exploding adimensional entity, mental and physical features may reasonably be assumed to have been present in the original entity, presumably in a potential state, and to have been transferred to newly born elementary particles. It follows that all entities generated during cosmic phylogenesis (from quarks to organisms and beyond) owe their physical and mental traits to the energy fields of elementary particles and of the more complex entities they generated by fusing complementary fields. The latter entities are assumed to be initially endowed with simple material bodies and primitive immaterial minds, and to be displaying bodies and minds of progressively greater complexity with evolving phylogenetic times.

This simple outline of evolutionary progress is in agreement with the utterly undifferentiated nature of the first big bang products. Sharing the energy fields of complementary particles/entities is likely to have been the mechanism generating more complex entities. Conversely, non-complementary entities were prevented from generating more complex assemblies because of their incompatible nature (due to bearing the same electric charge or other discrepancies) or the annihilation of the resulting products (antiparticles). This postulated mechanism implies that the architecture of properly assembled entities not only allowed their survival and permanence but was also instrumental in fostering the generation of higher order entities with more complex energy fields, degrees of freedom and mental capacities. Conversely, assembled units incapable of interacting with other entities were bound to remain locked in at their own level (evolutionary dead ends) as bodies that are still widely distributed in the universe.

At each level of organization, the original properties of constituent subunits are modulated by their inclusion in a more complex system. A helpful example is provided by the aromatic carbon compounds whose π electrons derived from the constituent carbon atoms are distributed over the entire aromatic ring. This feature shows that integration of constituent parts into a more complex entity requires a fraction of the individual energies to be shared by the novel unit in order to support its features. Could the sharing of energy from the constituent parts be considered a generalized event accounting for the genesis of simple and more complex levels of organization?

It seems obvious that the very existence of any unit made up of parts implies the coexistence of an energetic structure that prevents its dissolution by keeping parts together. Any such system should also be assumed to be potentially capable of associating with other units with similar or different levels of complexity by virtue of energetic sharing. One may then entertain the hypothesis that in a system of systems, energetic exchanges may not only take place between units of the same organization level, but also between more complex and less complex levels. In other words, one may ask whether energetic links exist between contiguous and distant levels of integration.

The entire field of chemistry provides a positive answer to this question. Chemical bonds which hold molecules arise from the energetic structure of atoms, just as bonds holding atoms together arise from the energetic fields of subatomic components. Likewise, more subtle, diverse and evanescent bonds support the existence of more complex systems like cells, organisms, and ecological systems. At each level, proper bonds arise from components. In brief, the very concept of integration implies that the dynamic cohesion of any system is based on energy flowing within the entire unit.

This concept may be generalized by stating that, irrespective of the level of complexity, each integrated unit owes its existence to the energetic structure generated by constituent units and subunits. Such a principle is validated by relatively simple constructs as well as by more complex systems encompassing more levels of integration. It follows that even the topmost level is assumed to share part of the energy arising from the lowest level, unless reasons are given to imply or suspect the presence of unlikely discontinuities in the energetic framework. Hence, humans may be presumed to share energies arising from less complex levels, down to elementary particles and mass-free energy, and to contribute energy to the higher systems they belong to, such as family, corporation, society, etc.

By analogy to ourselves, we may incidentally use the term 'soul' to denote the energetic structure responsible for the dynamic cohesion of any such system. From this point of view, π electrons may be considered the 'soul' of aromatic carbon compounds. Energy fields are known to extend to sidereal distances or to be limited to infinitesimal spaces, just as the size of the indivisible material 'atom' has progressively been contracted to reach the configuration of elementary particles or strings. Therefore, we find ourselves in agreement with Bertrand Russell, *"matter is a convenient concept to describe what is occurring where there is no real matter,"* and with Albert Einstein, *"we may thus consider matter as insisting in space regions where field is strikingly intense ... In this novel type of physics, field and matter cannot coexist, as field is the only reality."*

(f) Last but not least, the mind is responsible for the so-called paranormal events including telepathy, precognition, telekinesis, white and black magic, mystical states and transcendental meditation (Dunne & Jahn, 2005; Giuditta, 2010). They imply a capacity to overcome temporal boundaries and spatial distances, as mutual interactions occur between objects and organisms separated by space and time. The existence of paranormal phenomena is supported by an overwhelming literature, and by a multitude of human experiences common to all societies and cultures, accumulated throughout centuries under a variety of climates. Some of us have also experienced them. If miracles and interventions of the divine providence are to be included among such phenomena since their existence is accepted by dogma of the Catholic church, one should ask where the boundary with paranormal events is placed.

It is no mystery that the existence of these phenomena has received little or no credence by official science. Nonetheless, one should remember, *"there are more things in heaven and earth that can be dreamt of by thy philosophy"* (Shakespeare). Even official science cannot claim to be the only judge of truth and reality.

Conclusion

How do I conceive the nature of mind? Simply, by identifying it with the energy fields of elementary particles and related dynamic structures resulting from the progressive evolution of systems and systems of systems. By incorporating elementary particles into atoms, molecules, primitive cells and multicellular organisms, and finally attaining the astonishing structure of the human brain, it may not be unreasonable to expect that complex energy fields subsuming qualities and capacities of the human mind may have evolved from the *primordial* energy. This way of thinking implies a unitary view of the universe, and the belief that mind and matter are two sides of the same coin. Mind is assumed to be fully pervasive (panpsychism) but hardly perceived by sensory organs that only respond to the superficial appearance of bodies.

My thoughts on the nature of mind are to be regarded as plausible as the ephemeral figures drawn by the flocks of birds flying over Italian cities. These changing and living clouds undergo never-ending fragmentation and restructuring, governed by a collective logic which is hard to understand but is a metaphor for freedom. Thoughts on the nature of mind should be as free as the flights of birds. They should avoid the boundaries of scientific or theological dogmas, and should make good use of the precious humility provided by the freedom of being wrong in the quest for truth.

The mind we seek is the beloved child of the spirit, and spirit blows where it likes, more than the wind. Hence, how can we hope to succeed in understanding the nature of mind with rigid formulations and subtle distinctions, or by cherishing preconceived ideas in which we might remain entrapped, like birds in a net? Perhaps, the best way to get close to the nature of mind, which is now holding these thoughts but a moment ago was yielding different images, might be to try and describe its multifarious mental processes in the following somewhat poetic metaphor: the elegant turning of birds, flying free in the sky, majestically joining in large multitudes, and happily indifferent to sudden mutual departures.

*

The propositions listed above, if accepted, may lead to the following alternative theories:

(i) If mental features are believed to be present in prokaryotes, it follows that the human mind has evolved in parallel with the evolution of the human body;

(ii) If mental features are believed to be present in elementary particles, it follows that the human mind has evolved since the beginning of the universe, after traversing prebiotic and biological segments of cosmic evolution.

Consistent with (ii), the energy fields of elementary particles are proposed to be responsible for the evolution of material and mental properties and capacities, for the relative dynamic stability of material bodies and for the immaterial nature of mind.

Endnotes

Ambrosini, M. V., & Giuditta, A. (2001)." Learning and sleep: The sequential hypothesis." *Sleep Med.Rev.* 5:477–490.

Baluška, F., & Mancuso, S. (2009). "Deep evolutionary origins of neurobiology." *Commun. Integrat. Biol.* 2:1–6.

Darwin, C. (2010). *Taccuini filosofici*. Attanasio A., ed. 233.

Dunne, B. J., & Jahn, R. G. (2005). "Consciousness, information, and living systems." *Cell. Mol. Biol.* 51:703–714.

Ehrenfeucht, A., Prescott, D. M., & Rozenberg, G. (2007). "A model for the origin of internal eliminated segments (IESs) and gene rearrangement in stichotrichous ciliates." *J. Theor. Biol.* 244:108–114.

Giuditta, A. (2004). "Essay on the nature of mind." *Riv. Biol.* 97:187–196.

Giuditta, A. (2008). "Natural computing and biological evolution: A new paradigm." *Riv. Biol.* 101:119–128.

Giuditta, A. (2010). "The 1907 psychokinetic experiments by prof. Filippo Bottazzi." *J. Sci. Exploration*, 24:495–512.

Giuditta, A. (2011). "Sull'origine filogenetica della mente umana." *Rend. Acc. Sc. Fis. Mat. Napoli* LXXVIII: 345–350.

Tagkopoulos, I., Liu, Y. C., & Tavazoie, S. (2008). "Predictive behavior within microbial genetic networks." *Science* 320:1313–1317.

Wagner, U., Gais, S., Haider, H., Verleger, R., & Born, J. (2004). "Sleep inspires insight". *Nature* 427:352–355.

Evolution and Formative Causation

RUPERT SHELDRAKE

"Like father, like son" was a proverb in the Middle Ages; the Latin version "*quails pater talis filus*," played the same role in ancient Rome. The general principles of heredity had been known all over the world for millennia: children generally resemble their parents; they are usually more like members of their immediate family than unrelated people. It was also common knowledge that the same principles apply to animals and plants. Long before Darwin's theory of evolution and the pioneering genetic research of Gregor Mendel, people were breeding plants and animals selectively, creating an astonishing array of domesticated varieties, like dogs from Afghan hounds to Pekinese; and cabbages from broccoli to kale.

The discoveries of Mendel and Darwin were based on the practical successes of many generations of farmers and breeders. Darwin studied the subject for years. He subscribed to specialist publications like *Poultry Chronicle* and the *Gooseberry Growers' Register*, and grew fifty-four varieties of gooseberry in his garden at Down House, in Kent. He drew on the experience of cat and rabbit fanciers, horse and dog breeders, bee-keepers, horticulturalists and farmers. He joined two of the London pigeon clubs, visited fanciers to see their birds, and kept all the breeds he could obtain. He summarized this wealth of information in his book *The Variation of Animals and Plants Under Domestication*, which is one of my favorite books on biology.[1] The power of selective breeding suggested that a similar process worked spontaneously in the wild: natural selection.

Genetics is now at the very centre of biology. The standard view is that hereditary information is coded in the genes. The words "hereditary" and "genetic" are treated as synonyms. After the discovery of DNA in 1953, the nature of heredity appeared to be fully understood in molecular terms, at least in principle. The human genome project, completed in the year 2000, was a culminating technical triumph.

From a materialist point of view, non-material inheritance is impossible, except for cultural inheritance. Everyone agrees that cultural inheritance—say, through language—involves a transfer of information that is not genetic. But all other forms of inheritance *must* be material: there is no other possibility.

Yet several forms of material inheritance are known to be non-genetic. Cells inherit patterns of cell organization and structures like mitochondria directly from their mother cells, not through genes in the cell nuclei. This non-nuclear inheritance is called cytoplasmic inheritance. Animals and plants are also influenced by characteristics acquired by their ancestors. An inheritance of acquired characteristics can take place epigenetically, as opposed to genetically, through chemical changes that do not affect the underlying genetic code.

The unfamiliar idea of the non-material transmission of form and organization used to be the mainstream view and twentieth-century genetics developed in reaction against it. But even materialists end up with non-material explanations.

In the ancient world, almost no one believed that the form of an acanthus plant or a hawk was inherited through seeds or eggs alone. Platonists thought plants and animals were somehow shaped by the transcendent Idea or Form of their species. Modern Platonists, like René Thom, agree. They see the ideal Form of a species as a mathematical structure or model that is "reified" in physical plants or animals. The mathematical model for an acanthus plant is not embedded in the genes: it exists in a mathematical realm that transcends space and time. Human mathematical models are mere approximations to these mathematical archetypes.

Aristotle, Plato's student, disagreed. The forms of the species were not outside space and time, but inside space and time. They were *immanent*, meaning "dwelling in," not transcendent meaning "climbing beyond." Instead of an archetype in a *transcendent* mind-like realm, the form of the body was in the soul, which attracted the developing animal or plant towards its final form. The soul served both as its formal cause, the cause of the body's form, and it final cause, the end or goal towards which the organism was attracted.

In the European Middle Ages, Aristotle's theory, as modified and interpreted by Thomas Aquinas, was the basis of the orthodox understanding of causation. A process of change, like the growth of a walnut tree from a nut, involved four kinds of cause. The material cause was the matter out of which the plant was made, the nut and the matter it took up from its surroundings as it grew, like water and minerals from the soil. The moving cause was the energy that powered it, from sunlight. The formal cause was the cause of the form, or structure, the walnut-tree form in the plant's soul. The final cause was the goal or purpose of the plant's growth, namely the mature tree producing nuts to reproduce itself.

An architectural analogy provides another way of thinking about the four causes. In order to build a house, there must be building materials, like bricks and cement. These are the material causes. Putting them in the right places requires the energy of the builders and their machinery: these are the moving causes. The places in which they put the materials are specified by the architect's plan: this is the formal cause. All this activity is happening because the person paying for the house wants to live in it; this is the purpose or final cause. All four causes are necessary: the house would not exist without the materials of which it is made, or the energy of the builders, or a plan, or a motivation for building it. In living organisms, immaterial souls provide both plans and purposes.

An essential feature of the mechanistic revolution in the seventeenth century was the abolition of souls, along with formal and final causes. Everything was to be explained mechanistically in terms of material and moving causes. This meant that the source of an organism's form must already be present inside the fertilized egg as a material structure.

From the seventeenth century until the beginning of the twentieth, biologists were divided into two main camps: the mechanists and the vitalists. Both needed to explain heredity. The vitalists continued the Aristotelian tradition: organisms were shaped by souls or non-material vital forces. The problem was that they could not see how these non-material forces worked or how they interacted with bodies.

The mechanists preferred a material explanation, but they too soon ran into problems. To start with, they proposed that the animals and plants were already present in the fertilized egg, in a miniature form. They were *pre-formed*. Development was a growth and unfolding—or inflation—of these pre-formed material structures. A few pre-formationists believed that the tiny unexpanded organisms came from eggs, but most thought they were in sperm, and some claimed to have proved it. One microscopist saw miniature horses in horse sperm, and miniature donkeys in donkey sperm, with big ears. Another saw tiny homunculi in human sperm.

Although pre-formationism was easy to understand and apparently supported by microscopic evidence, it ran into serious theoretical difficulties in relation to the succession of generations. As their vitalist opponents pointed out, if a rabbit grows from a miniature rabbit in a fertilized egg, the tiny rabbit in the egg must contain even tinier rabbits in its gonads, and so on *ad infinitum.*[2]

Pre-formationism was finally refuted in the late eighteenth century. As researchers looked at developing embryos in detail, they found that new structures appeared that were not there before. For example, the intestine, formed by the infolding of a sheet of tissue from the ventral surface, produced a gutter which in time transitioned itself into a closed tube.[3] By the mid-nineteenth century the evidence was overwhelming: development involved the formation of new structures that were not already present. Development was *epigenetic*, from the Greek *epi*, over and above, and *genesis*, organization. New structures appeared that were not already present in the egg. Epigenesis supported both Platonic and Aristotelian schools of thought. Neither supposed that all of an organism's form was contained in the matter of the fertilized egg. Its form was derived from a Platonic Idea or a soul.

By contrast, mechanists faced the daunting challenge of explaining how more material form could arise from less and develop in a highly ordered way. In the 1880s, August Weismann (1834–1914) thought he had found the answer. He made a theoretical division of organisms into two parts, the body, or somatoplam, and the germ-plasm, a material structure present in the fertilized egg. He thought the germ-plasm was an active agency, containing "determinants" that shaped the form of the somatoplasm. The germ-plasm affected the somatoplasm but not vice versa. The determinants "directed" the formation of the adult organism, but the germ-plasm itself was passed on unchanged through eggs and sperm.

By the middle of the twentieth century, the discovery of genes located in chromosomes inside cell nuclei seemed to have confirmed Weismann's theory. The genes were the germ-plasm, replicated more or less unchanged in every cell division. The discovery of the structure of the genetic material, DNA, and the cracking of the genetic code in the

1950s showed how the Weismann doctrine could be reduced to the molecular level. DNA was the germ-plasm, proteins were the somatoplasm. The DNA coded for the structure of proteins, but not vice versa. Francis Crick called this the "central dogma of molecular biology." Meanwhile, the neo-Darwinian theory explained evolution in terms of random mutations in genes and changes in gene frequencies in populations as a result of natural selection. The triumphs of molecular genetics combined with the neo-Darwinian theory of evolution seemed to provide overwhelming evidence for the material theory of inheritance. But this triumph was more a matter of rhetoric than reality.

DNA molecules are molecules. They are not "determinants" of particular structures, even though biologists often speak of genes "for" structures or activities, such as genes "for" curly hair or "for" nest-building behavior in sparrows. Genes are not selfish and ruthless, as if they contained gangster homunculi. Nor are they plans or instructions for organisms. They merely code for the sequences of amino acids in protein molecules.

If genetic programs were carried in the genes, then all the cells of the body would be programmed identically, because in general they contain exactly the same genes. Your limbs contain exactly the same kind of protein molecules, as well as chemically identical bone, cartilage and nerves. Yet arms and legs have different shapes. Clearly, the genes alone cannot explain these differences. They must depend on formative influences that act differently in different organs and tissues as they develop. These influences cannot be inside the genes: they extend over entire tissues and organs. At this stage, in most conventional explanations, the concept of the genetic program fades out, and is replaced by vague statement about "complex spatio-temporal patterns of physico-chemical activity not yet fully understood" or "mechanisms as yet obscure" or "chains of parallel and successive operations that build complexity."[4]

In spite of the fact that many biologists now recognize that it is misleading, the genetic program continues to play a large conceptual role in modern biology. There seems to be a need for such an idea. Mechanistic biology grew up in opposition to vitalism. It defined itself by denying that living organisms are organized by purposive, mind-like principles, but then reinvented them in the guise of genetic programs and selfish genes.

The dominant paradigm of modern biology, although nominally mechanistic, is remarkably similar to vitalism, with "programs" or "information" or "instructions" or "messages" playing the role formerly attributed to souls.

In the wake of the Human Genome Project, the mood changed dramatically. The optimism that life would be understood if molecular biologists knew the "programs" of an organism gave way to the realization that there is a huge gap between gene sequences and actual human beings.

One of the biggest controversies in twentieth-century biology concerned the inheritance of acquired characteristics, the ability of animals and plants to inherit adaptations acquired by their ancestors. In Darwin's day, most people assumed that acquired characteristics could indeed be inherited. Jean-Baptiste Lamarck had taken this for granted in his theory of evolution published more than fifty years before Darwin's, and the inheritance of acquired characteristics was often referred to as "Lamarckian inheritance." Darwin shared this belief and cited many examples to support it.[5] In this respect Darwin was a Lamarckian, not so much because of Lamarck's influence but because he and Lamarck both accepted the inheritance of acquired characteristics as a matter of common sense.[6]

Lamarck placed a strong emphasis on the role of behavior in evolution: animals' development of new habits in response to needs led to the use or disuse of organs, which were accordingly either strengthened or weakened. Over a period of generations, this process led to structural changes that became increasingly hereditary. Lamarck's most famous example was the giraffe, whose long neck was acquired through the habit of stretching up to eat the leaves of trees over many generations. In this respect too, Darwin agreed with Lamarck, and he provided various illustrations of the hereditary effects of the habits of life. For example, ostriches, he suggested, may have the lost the power of flight through disuse and gained stronger legs through increased use over successive generations.[31] Darwin was very conscious of the power of habit, which was for him almost another name for nature. Francis Huxley summarized Darwin's attitude:

A structure to him meant a habit, and a habit implied not only
an internal need but outer forces to which, for good or evil, the

organism had to become habituated... In one sense, therefore, he might well have called his book *The Origin of Habits* rather than *The Origin of Species.*

The problem was that no one knew how acquired characteristics could be inherited. Darwin tried to explain it with his hypothesis of "pangenesis." He proposed that all the units of the body threw off tiny "gemmules" of "formative matter," which were dispersed throughout the body and aggregated in the buds of plants and in the germ cells of animals, through which they were transmitted to the offspring.[33]

The neo-Darwinian theory of evolution, which became orthodox in the West in the twentieth century, differed from the Darwinian theory in denying the inheritance of acquired characteristics in favor of genes. Lamarckian inheritance was treated as a heresy. By contrast, in the Soviet Union the inheritance of acquired characteristics became the orthodox doctrine from the 1930s to the 1960s.

The taboo against the inheritance of acquired characteristics began to dissolve around the turn of the millennium, with a growing recognition that some acquired characteristics can indeed be inherited. This kind of inheritance is now called "epigenetic inheritance." In this context, the word "epigenetic" signifies "over and above genetics." Although epigenetic inheritance breaks the taboo against the inheritance of acquired characteristics, it does not challenge the materialist assumption that heredity is material; it is another kind of material inheritance. It affects which genes are "switched on" or "switched off." and consequently affects what proteins a cell makes. But genes and proteins cannot in themselves explain morphogenesis or instinctive behavior.

The only way of making sense of inherited patterns or organization is in terms of top-down causation by higher-level patterns, or "systems properties," or fields. My own hypothesis, developed in detail in *A New Science of Life: The Hypothesis of Formative Causation,*[7] is that the formation of habits depends on a process called morphic resonance. Similar patterns of activity resonate across time and space with subsequent patterns. This hypothesis applies to all self-organizing systems, including atoms, molecules, crystals, cells, plants, animals and animal societies. All draw upon a collective memory and in turn contribute to it.

A growing crystal of copper sulphate, for example, is in resonance with countless previous crystals of copper sulphate, and follows the same habits of crystal organization, the same lattice structure. A growing oak seedling follows the habits of growth and developments of previous oaks. When an orb-web spider starts spinning its web, it follows the habits of countless ancestors, resonating with them directly across space and time. The more people who learn a new skill, such as snowboarding, the easier it will be for others to learn it because of morphic resonance from previous snowboarders.

The hypothesis of formative causation proposes that morphogenetic fields play a causal role in the development and maintenance of the forms of systems at all levels of complexity. In this context, the word "form" is taken to include not only the shape of the outer surface or boundary of a system, but also its internal structure. This suggested causation of form by morphogenetic fields is called formative causation in order to distinguish it from the energetic type of causation with which physics already deals so thoroughly. For although morphogenetic fields can only bring about their effects in conjunction with energetic processes, they are not in themselves energetic.

Morphogenetic fields can be regarded as analogous to the known fields of physics in that they are capable of ordering physical changes, even though they themselves cannot be observed directly. Gravitational and electromagnetic fields are spatial structures which are invisible, intangible, inaudible, tasteless and odorless; they are detectable only through their respective gravitational and electromagnetic effects. In order to account for the fact that physical systems influence each other at a distance without any apparent material connection between them, these hypothetical fields are endowed with the property of traversing empty space, or even actually constituting it. In one sense, they are non-material; but in another sense, they are aspects of matter because they can only be known through their effects on material systems. In effect, the scientific definition of matter has simply been widened to take them into account. Similarly, morphogenetic fields are spatial structures detectable only through their morphogenetic effects on material systems; they too can be regarded as aspects of matter if the definition of matter is widened still further to include them.

The idea of a process whereby the forms of previous systems influence the morphogenesis of subsequent similar systems is difficult to express in terms of existing concepts. The only way to proceed is by means of analogy. The physical analogy which seems most appropriate is that of *resonance*. Energetic resonance occurs when a system is acted on by an alternating force which coincides with its natural frequency of vibration. Examples include the "sympathetic" vibration of stretched strings in response to appropriate sound waves; the tuning of radio sets to the frequency of radio waves given out by transmitters; the absorption of light waves of particular frequencies by atoms and molecules, resulting in their characteristic absorption spectra; and the response of electrons and atomic nuclei in the presence of magnetic fields to electromagnetic radiation in Electronic Spin Resonance and Nuclear Magnetic Resonance. Common to all these types of resonance is the principle of selectivity; out of a mixture of vibrations, however complicated, the systems respond only to those of particular frequencies.

A "resonant" effect of form upon form across space and time would resemble energetic resonance in its selectivity, but it could not be accounted for in terms of any of the known types of resonance, nor would it involve a transmission of energy. In order to distinguish it from energetic resonance, this process will be called *morphic resonance.*

Morphic resonance is analogous to energetic resonance in a further respect: it takes place between vibrating systems. Atoms, molecules, crystals, organelles, cells, tissues, organs and organisms are all made up of parts in ceaseless oscillation, and all have their own characteristic patterns of vibration and internal rhythm; the morphic units are dynamic, not static. But whereas energetic resonance depends only on the specificity of response to particular frequencies, to "one-dimensional" stimuli, morphic resonance depends on three-dimensional patterns of vibration. What is being suggested here is that by morphic resonance the form of a system, including its characteristic internal structure and vibrational frequencies, becomes present to a subsequent system with a similar form; the spatio-temporal patterns of the former *superimposes* itself on the latter.

Morphic resonance takes place through morphogenetic fields and indeed gives rise to their characteristic structures. Not only does a specific morphogenetic field influence the form of a system, but also the form of this system influences the morphogenetic field and through it becomes present to subsequent similar systems.

To summarize briefly, the morphic resonance hypothesis proposes that:

1. Self-organizing systems including molecules, cells, tissues, organs, organisms, societies and minds are made up of nested hierarchies of holons or morphic units. At each level the whole is more than the sum of its parts, and these parts themselves are wholes made up of parts.

2. The wholeness of each level depends on an organizing field, called a morphic field. This field is within and around the system it organizes, and is a vibratory pattern of activity that interacts with electromagnetic and quantum fields of the system. The generic name "morphic field" includes

(a) Morphogenetic fields that shape the development of plants and animals.

(b) Behavioral and perceptual fields that organize the movements, fixed-action patterns and instincts of animals.

(c) Social fields that link together and co-ordinate the behavior of social groups.

(d) Mental fields that underlie mental activities and shape the habits of minds.

3. Morphic fields contain attractors (goals), and chreodes (habitual pathways toward those goals) that guide a system towards its end state, and maintain its integrity, stabilizing it against disruptions.

4. Morphic fields are shaped by morphic resonance from all similar past systems, and thus contain a cumulative collective memory. Morphic resonance depends on similarity, and is not attenuated by distance in space or time. Morphic fields are local, within and around the systems they organize, but morphic resonance is nonlocal.

5. Morphic resonance involves a transfer of form or in-*form*-ation rather than a transfer of energy.

6. Morphic fields are fields of probability, like quantum fields, and they work by imposing patterns on otherwise random events in the system under their influence.

7. All self-organizing systems are influenced by self-resonance from their own past, which plays an essential role in maintaining a holon's identity and continuity.

This hypothesis leaves open the question of how morphic resonance actually works. There are several suggestions. One is that the transfer of information occurs through the "implicate order," as proposed by the physicist David Bohm.[8] The implicate or enfolded order gives rise to the world we can observe, the "explicate order," in which things are located in space and time. In the implicate order, according to Bohm, "everything is enfolded into everything. Or resonance may pass through the quantum-vacuum field, also known as the zero-point energy field, which mediates all quantum and electromagnetic processes."[9] Or similar systems might be connected through hidden extra dimensions, as in string theory and M-theory.[10] Or maybe it depends on new kinds of physics as yet unthought-of.

Morphic resonance is non-energetic, and morphogenetic fields themselves are neither a type of mass nor energy. Therefore, there seems to be no *a priori* reason why it should obey the laws that have been found to apply to the movements of bodies, particles and waves. In particular, it need not necessarily be attenuated by either spatial or temporal separation between similar systems; it could be just as effective over ten thousand miles as over a yard, and over a century as over an hour.

If morphic resonance is indeed unattenuated by time and space, it is not reasonable to assume that it takes place only from the past; that only morphic units which have already existed are able to exert a morphic influence in the present. The notion that *future* systems, which do not yet exist, might be able to exert a causal influence "backwards" in time may not seem worth considering. Most people assume that time-reversed causation is scientifically impossible. But, surprisingly, most of the laws of physics are reversible, and work just as well from the future to the past as from the past to the future. In James Clark Maxwell's classical equations for electromagnetic waves, put forward in 1864, there are two answers that describe the movement of light waves. In one answer, the waves move at the speed of light from the present to the future, as in the conventional understanding of causation. But in the other answer, the waves move from the present into the past at the speed of light, in the opposite direction to ordinary causation. These waves moving backward in time are called "advanced waves." They imply influences working backwards in time. Advanced waves are part of the mathematics of electromagnetism, but physicists ignore them because they are regarded as "non-physical."

Yet, in the "transactional" interpretation of quantum mechanics, quantum processes are seen as standing waves between emitters and absorbers, with forward-in-time waves moving from emitter to absorber, and backward-in-time waves from absorber to emitter.

Although a creative agency capable of giving rise to new forms and new patterns of behavior in the course of evolution would necessarily transcend individual organisms, it need not transcend all nature. It could, for instance, be immanent within life as a whole; in this case it would correspond to what Bergson called the *élan vital*.[11] Or it could be immanent within the planet as a whole, or the solar system, or the entire universe. There could indeed be a hierarchy of immanent creativities at all these levels.

Such creative agencies could give rise to new morphogenetic fields by a kind of causation very similar to a kind of conscious causation. In fact, if such creative fields are admitted at all, then it is difficult to avoid the conclusion that they must in some sense be conscious selves.

All religions assume that human consciousness plays an essential role in the world and in human destiny. Humans have the potential to participate in ultimate Being, or God, or cosmic consciousness, or divine life, or nirvana. Experiences of unity with a greater being, or mystical experiences, are surprisingly common. In addition, many thousands of people have reported near-death experiences, in most cases with life-changing effects.

The Jesuit biologist Teilhard de Chardin (1881–1955) thought that the entire evolutionary process was moving toward an end-point of "maximum organized complexity," which he called the Omega point. The Omega point was the attractor of the entire cosmic evolutionary process, and through it consciousness would be transformed.[12]

If the transformation of human consciousness is the goal of evolution, then why do there need to be a billion stars beside our sun in our galaxy and billions of other galaxies beyond it? Or is consciousness developing throughout the universe? And will our consciousness ultimately make contact with those other minds? These are open questions, and neither conventional science nor traditional religions have any ready answers. Philosophers like Teilhard de Chardin point to new possibilities that go beyond the speculations of scientists by seeing consciousness as central to the evolutionary process. But even for the most materialistic of scientists, consciousness has a privileged position as the matrix of human knowledge, the basis of science itself.

Endnotes

1 Darwin, C. (1875) *The Variation of Animals and Plants Under Domestication.* Murray, London.

2 Needham, J. (1959) *A History of Embryology. Cambridge University Press, Cambridge.*

3 Holder, N. (1981) "Regeneration and Compensatory Growth," *British Medical Bulletin, 37, 227-32.*

4 E.g. Carroll, S.B. (2005) *Endless Forms Most Beautiful, Quercus, London.*

5 See Darwin, C. (1859) *The Origin of Species, and (1875) The Variation of Animals and Plants Under Domestication.*

6 Mayr, E. (1982. *The Growth of Biological Thought, Harvard University Press, Cambridge, MA.*

7 Sheldrake, R. (1981) *A New Science of Life: The Hypothesis of Formative Causation. Los Angeles, CA, J.P. Tarcher.*

8 Bohm, D. (1980) *Wholeness and the Implicate Order, Routledge & Kegan Paul, London.*

9 Laszlo, E. (2007) *Science and the Akashic Field, Inner Traditions, Rochester, VT.*

10 Carr, B. (2008) "Worlds apart? Can psychical research bridge the gap between matter and minds?" *Proceedings of the Society for Psychical Research, 59,1-96.*

11 Bergson, H. (1911) *Creative Evoltion. Macmillan, London.*

12 De Chardin, P.T. (1959) *The Phenomenon of Man. Collins, London.*

A Syntropic Model of Consciousness

ULISSE DI CORPO AND ANTONELLA VANNINI[1]

Introduction

Consciousness, the "feeling of being alive," is still a mystery. Neuro-scientists tend to assume that consciousness emerges from matter, whereas quantum scientists tend to assume that matter emerges from consciousness.

In this paper, we suggest that consciousness is a property of a symmetric and complementary energy to physical energy. This hypothesis was first put forward by the mathematician Luigi Fantappiè and the paleontologist Pierre Teilhard de Chardin. Fantappiè noted that the fundamental equations which combine special relativity with quantum mechanics have a "physical" solution for energy that describes energy which diverges forwards in time and a "non-physical" solution which describes energy that diverges backward in time. The qualities of the non-physical energy match the mysterious qualities of life: energy concentration, increase in differentiation and complexity.

Since this energy propagates from the future, it is invisible to us. Whereas physical energy is visible and can be perceived, albeit indirectly, non-physical energy is invisible and can be felt.

The Unphysical Solution of Energy

The equation *E=mc2*, commonly associated with the work of Albert Einstein, was first published in 1890 by Oliver Heaviside[2] and then refined by Henri Poincaré[3] in 1900 and Olinto De Pretto[4] in 1903, who registered it at the *Regio Instituto di Scienze* and then published it in a paper together with the senator and astronomer Giovanni Schiaparelli. It seems that the Energy-Mass equation reached Einstein through his father Hermann, who was the owner of the *"Privilegiata Impresa Elettrica Einstein,"* working in the development of street lighting in Verona together with Olinto De Pretto. But the *E=mc²* had a major problem, it did not take into account motion, the momentum, which is also a form of energy.

Einstein solved the problem by adding the momentum and published in 1905, in his Special Relativity, the more complex energy/momentum/mass equation:

$$E^2 = p^2c^2 + m^2c^4$$

which relates energy (E), momentum (p) and mass (m).

This equation is a double order equation and has two solutions for energy: a positive time solution, which describes energy that diverges from the past to the future, and a negative time solution, which describes energy that diverges backward in time, from the future into the past. Since we move forward in time, the backward in time diverging energy turns into a converging, attractive force.

In 1905 energy flowing backward in time was considered impossible. Einstein solved the problem by removing the momentum from the equation and going back to the *E=mc²*, which has only a forward in time solution. He could do this since the speed of physical bodies is practically nil compared to the speed of light.

Everything worked fine until 1924, when Wolfgang Pauli discovered that the spin of subatomic particles (which is a momentum) nears the speed of light. Quantum mechanics thus requires the extended *energy/momentum/mass* equation, with its inconvenient backward in time solution!

The first equation that combines Special Relativity and Quantum Mechanics dates back to 1926 and was formulated by the physicists Klein and Gordon. This equation has two solutions: a backward in time (advanced waves) and a forward in time (delayed waves). The advanced wave solution was rejected, since it implies retrocausality, which was considered impossible.

The second equation was formulated in 1928 by Paul Dirac. Dirac was trying to solve the paradox of the backward in time energy solution, but he found the electron and the neg-electron (now named *positron*) that propagates backward in time. Positrons were observed experimentally in 1932 by Carl Anderson[5]. Shortly after, Pauli wrote an essay with the famous psychologist Carl Gustav Jung in which he posited that starting from the dual solution of the fundamental equations, we live in a super-causal world, with causes acting from the past and synchronicities acting from the future.

In 1933 Werner Heisenberg, who had a strong charismatic personality and a leading position in scientific institutions and academia, declared the backward in time solution impossible. From that moment, anyone who ventured into the study of the backward in time solution was discredited, losing academic credibility and the ability to publish and to talk at conferences.

Syntropy

Luigi Fantappiè had studied pure mathematics at the Normale di Pisa, the most exclusive Italian university, where he had been classmate of Enrico Fermi. Well known among physicists, in 1951 he was invited by Oppenheimer to become a member of the exclusive Institute for Advanced Study in Princeton and work directly with Einstein.

Fantappiè could not accept that Heisenberg had rejected half of the solutions of the fundamental equations in a totally subjective way. In 1941, while listing the properties of the forward and backward in time energy solutions, Fantappiè realized that the forward-in-time solution is governed by the law of *entropy* (the word entropy is the combination of the Greek words *en*=diverging and *tropos*=tendency), whereas the backward-in-time solution is governed by a symmetrical law that he named *syntropy* (from the combination of the Greek words *syn*=converging and *tropos*=tendency).

Entropy is the tendency towards energy dissipation, the famous second law of thermodynamics, also known as the law of heat death or disorder. To the contrary, syntropy is the tendency towards energy concentration, increase in differentiation, formation of structures, and organization.

Listing the mathematical properties of syntropy, Fantappiè recognized the mysterious qualities of life and in 1942 he wrote a booklet titled "The *Unified theory of the physical and biological world,*" in which he suggests that the physical/material world is governed by the law of entropy, whereas life is governed by attractors that retroact from the future and follow the law of syntropy.[6] The future is invisible and life constantly mediates the visible and physical universe with the invisible and immaterial universe of syntropy and vital energies.

In a letter to a friend Fantappiè described the discovery of syntropy with the following words:

> In the days just before Christmas 1941, as a consequence of conversations with two colleagues, a physicist and a biologist, I was suddenly projected in a new panorama, which radically changed the vision of science and of the Universe which I had inherited from my teachers, and which I had always considered the strong and certain ground on which to base my scientific investigations. Suddenly I saw the possibility of interpreting a wide range of solutions (the anticipated potentials) of the wave equation which can be considered the fundamental law of the Universe. These solutions had been always rejected as "impossible," but suddenly they appeared "possible," and they explained a new category of phenomena which I later named "syntropic," totally different from the entropic ones, of the mechanical, physical and chemical laws, which obey only the principle of classical causation and the law of entropy. Syntropic phenomena, which are instead represented by those strange solutions of the "anticipated potentials," should obey two opposite principles of finality (moved by a final cause placed in the future, and not by a cause which is placed in the past) and differentiation, and also non-causable in a laboratory. This last characteristic explains why this type of phenomena has never been reproduced in a laboratory, and its finalistic properties justified the refusal among scientists, who accepted without any doubt the assumption that finalism is a "metaphysical" principle, outside Science and Nature. This assumption obstructed the way to a calm investigation of the real existence of this second type of phenomena; an investigation which I accepted to carry out, even though I felt as if I were falling in an abyss, with incredible consequences and conclusions. It suddenly seemed as if the sky were falling apart, or at least the certainties on which

mechanical science had based its assumptions. It appeared to me clear that these "syntropic," finalistic phenomena which lead to differentiation and could not be reproduced in a laboratory, were real, and existed in nature, as I could recognize them in the living systems. The properties of this new law, opened consequences which were just incredible and which could deeply change the biological, medical, psychological, and social sciences.

The Invisible Side of Reality

We continuously experience forces and entities that we cannot observe directly but which exist objectively, independently of any human perception. One such force is gravity.

Suppose we hold a small object like a pencil between our thumb and forefinger and then release it. We observe that it falls to the floor and we say that the force of gravity causes it to fall. But do we actually see any downward force acting upon the pencil, something pulling or pushing it? Clearly not. We do not observe the force of gravity at all. Rather we deduce the existence of some unseen force (called gravity) acting upon unsupported objects in order to explain their otherwise inexplicable downward movement.

According to the energy/momentum/mass equation, half of the forces acting in the universe are entropic (visible) and half are syntropic (invisible) and nothing takes place without the interplay of both these forces. Gravity is described as a backward-in-time diverging force. But, since we move forward in time, this backward diverging force is for us a forward converging force.

Equations show that forward diverging forces cannot exceed the speed of light, whereas backward diverging forces can never propagate at speeds slower than that of light. Consequently, if the entropy/syntropy hypothesis is correct, we should observe that gravity propagates at an instantaneous speed.

But can we perform experiments in order to measure the speed of propagation of gravity? The answer has been provided by Tom van Flandern (1940–2009), an American astronomer specialized in celestial mechanics. Van Flandern noted that no aberration is observed when measuring gravity and that this puts the propagation of gravity at a speed higher than 10^{10} the speed of light.[7,8,9] With light the aberration is due to its limited speed. For example, light from the Sun requires about 500 seconds to travel to Earth. So when it arrives, we see the Sun in the sky in the position it actually occupied 500 seconds ago rather than in its present position. This difference amounts to about 20 seconds of arc, a large and noticeable amount to astronomers. From our perspective, the Earth is standing still and the Sun is moving. So it seems natural that we see the Sun where it was 500 seconds ago, when it emitted the light now arriving. Consequently, the light from the Sun strikes the Earth from a slightly displaced angle and this displacement is called aberration. Light aberration is due entirely to the finite speed of light.

If gravity would propagate with a finite speed we would expect gravity aberration. The Sun's gravity should appear to emanate from the position the Sun occupied when the gravity now arriving left the Sun. The Earth should "run into" the gravitational force, making it appear to come from a slightly displaced angle equal to the ratio of the Earth's orbital speed to the speed of gravity propagation.

Yet observations indicate that none of this happens in the case of gravity! There is no detectable delay for the propagation of gravity from Sun to Earth. The direction of the Sun's gravitational force is toward its true, instantaneous position, not toward a delayed position, to the full accuracy of observations. Gravity has no perceptible aberration and this tells that it propagates with infinite speed.

Van Flandern notes that gravity has some curious properties:

- One of them is that its effect on a body is apparently completely independent of the mass of the affected body. As a result, heavy and light bodies fall in a gravitational field with equal acceleration.
- Another is the seemingly infinite range of gravitational force. Truly infinite range is not possible when forces are conveyed forward-in-time.
- The other curious property of gravity is its instantaneous action and propagation, which can be explained only if we accept that gravity is a backward-in-time diverging force.

Complementarity

The energy/momentum/mass relation shows that the visible/diverging reality and the invisible/converging reality are united by the same equation. We can write, for example:

$$E_{total} = E_{visible} + E_{invisible}$$

The total energy is the sum of visible and invisible energy. Reality is visible and invisible. The visible reality expands and is governed by the law of entropy, whereas the invisible reality contracts and is governed by the law of syntropy.

We can write:

$$E_{total} = E_{entropic} + E_{syntropic}$$

The first law of thermodynamics states that energy is constant, since it cannot be created or destroyed, but only transformed. We can therefore replace energy with the number 1 and write:

$$1 = Entropy + Syntropy$$
$$Entropy = 1 - Syntropy$$

$$Syntropy = 1 - Entropy$$

These expressions show that entropy and syntropy are complementary parts of the same unity.[10] They also show that syntropy is profoundly different from negentropy. Syntropy and negentropy should not be confused together, since negentropy is defined as the opposite of entropy:

$$negentropy = -entropy$$

To the contrary, syntropy is defined as the complement to entropy.

Duality

The description of two complementary forces, one diverging and one converging, one visible and one invisible, one destructive and one constructive, can be found in many philosophies and religions. For example, in the Taoist philosophy all aspects of the universe are described as the interplay of two complementary and fundamental forces: the yang principle which is diverging, and the yin principle which is converging.

These two forces are part of a unity. In the visible side of reality, when one increases the other decreases, but as a whole their balance remains unchanged. This law is masterfully represented in the Taijitu symbol that is the union of these opposite forces, the yin and the yang, the diverging and converging forces whose combined action moves the universe in all its aspects: the sexes, seasons, day and night, life and death, full and empty, movement and repose, push and pull, dry and wet. Water takes on yang in its steaming form and yin in its icy form. Within the yin there is yang, and within the yang there is yin.

In the Taijitu symbol the yang principle is represented by the white color and has entropic properties, whereas the yin principle is represented by the black color and has syntropic properties. The Taijitu is a wheel that rotates constantly, changing the proportion of yin and yang (syntropy and entropy) in the visible and the invisible sides of reality. The Taijitu shows that a property of complementarity is that opposites attract each other. This property is well known in physics, but it is also true at the human level where people of opposite polarities are attracted to each other, as in males and females. Since the balance of these opposite forces remains unchanged, the Taoist philosophy suggests that the aim is to harmonize the opposites, thus creating unity.

Taijitu symbol

In Hinduism, the law of complementarity is described by the dance of Shiva and Shakti, where Shakti is the personification of the female principle and Shiva of the male principle. They represent the primordial cosmic energy and the dynamic forces that are thought to move through the entire universe. Shiva has the properties of the law of syntropy, whereas Shakti has the properties of the law of entropy and they are

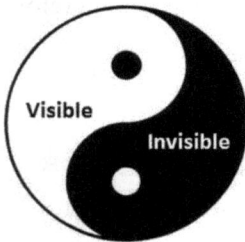

constantly combined together in an endless cosmic dance. Shakti can never exist apart from Shiva or act independently of him, just as Shiva remains a mere corpse without Shakti. All the matter and energy of the universe results from the dance of the two opposite forces of Shiva and Shakti. Shiva absorbs Shakti energy, turning it into a body and absolute pure consciousness, the light of knowledge. According to Hinduism, intelligence comes from the future (Shiva), whereas fearsomeness, ferocity, and aggression come from the past (Shakti). Shakti is the energy of the physical and visible world whereas Shiva is the consciousness that transcends the visible world. Each aspect of Shiva has a Shakti component, however, linked to the physical world. The evolution of this endless dance between Shakti and Shiva has the function of bringing life towards Unity.

In the psychological literature of the 20th century Carl Jung added synchronicities (i.e. syntropy) to causality (i.e. entropy). According to Jung, synchronicities are the experience of two or more events that are apparently causally unrelated or unlikely to occur together by chance, yet they are experienced as occurring together in a meaningful manner.

The concept of synchronicity was first described in this terminology by Jung in the 1920s. The concept does not question, or compete with, the notion of causality. Instead, it maintains that just as events may be grouped by causes, they may also be grouped by finalities, a meaningful principle. He coined the word "synchronicity" to describe what he called "temporally coincident occurrences of acausal events." He described synchronicity variously as an "acausal connecting principle," "meaningful coincidence," and "acausal parallelism."

Jung gave a full statement of this concept in 1951 when he published the paper *Synchronicity—An Acausal Connecting Principle*, jointly with a related study by physicist Wolfgang Pauli.[11] In Jung's and Pauli's description, causality acts from the past, whereas synchronicity acts from the future. Synchronicities are meaningful since they lead towards a finality, providing a direction to events which correlates them in apparently acausal ways.

Jung and Pauli described causality and synchronicity as acting on the same indestructible energy. They are united by this energy, but at the same time they are complementary.

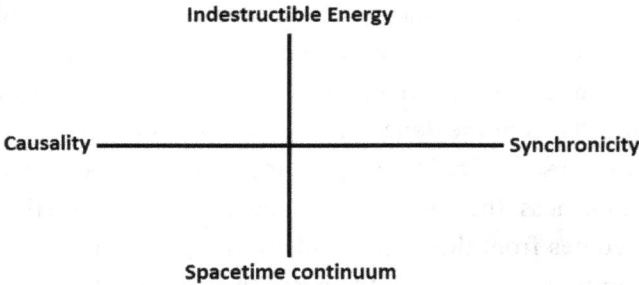

This game between entropy and syntropy can also be clearly seen in metabolism. Syntropy concentrates energy in ever smaller spaces increasing order and organization, but since the concentration of energy cannot increase indefinitely, at some point the system releases energy and matter, thus activating the opposite process of entropy and an exchange of energy and matter with the environment. Exchange between life and the environment results in a continuous process of construction and destruction that allows life to evolve. Exchange reveals the principle of complementarity, which is a fundamental property of life at all its levels of organization, from the organic/biological level to economics.

In metabolism, *Entropy* corresponds to *Catabolic* processes, which transform higher level structures into lower level structures with the release of energy in the form of chemical energy (ATP) and thermal energy, and *Syntropy* corresponds to *Anabolic* processes, which transform simple structures into complex structures, for example nutritive elements into bio-molecules, with the absorption of energy.

In the entropy/syntropy hypothesis, complementarity is represented as a see-saw where entropy and syntropy play at the opposite sides. Life is at the middle.

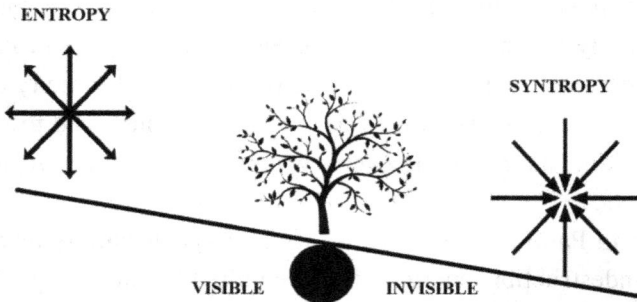

Entropy and Syntropy constantly playing, transforming energy

This representation clearly shows that when entropy goes down syntropy rises and when entropy rises syntropy goes down. It also shows that we can act on the invisible side of reality, just by reducing or increasing the entropy level of the visible plane. If we want to reduce syntropy we increase entropy, if we want to increase syntropy we need to decrease entropy.

The Feeling of Life

Wheeler, Feynman[12] and Fantappiè showed that advanced waves (i.e. syntropy) behave as absorbers whereas delayed waves (i.e. entropy) behave as emitters. Fantappiè adds to this description that living systems, as a consequence of the fact that they are syntropic, are energy absorbers and the energy balance is therefore always positive, in favor of absorption. The assertion that living systems absorb energy is consistent with the fact that nearly all the energy used by humanity derives from biological masses: wood, coal, petrol, gas, and bio-fuels.

The distinction between absorbers and emitters provides an interesting insight into one of the basic property of life: the *"feeling of life."* According to Damasio, the *"background feeling,"* which is the equivalent of the *"feeling of life"* is the fundamental element of consciousness and life.[13]

This background feeling of absorbing energy can be considered the essence of life itself and of the feeling of life. If this is correct, it would follow that the feeling of life, consciousness, is a direct consequence of advanced waves.

The equivalence *"feeling of life = advanced waves"* leads to the assumption that systems based on the positive energy solution (entropy), as for example machines and computers, will never be endowed with the *"feeling of life"* independently from their complexity, whereas systems based on the negative energy solution (syntropy), as for example life itself, should always have a *"feeling of life"* independently of their complexity.

Anticipatory Systems

Pre-stimuli activations seem to play a key role in the survival and welfare of all living systems. Robert Rosen (1934–1998), a theoretical biologist and professor of biophysics at the Dalhousie University, coined the expression *"Anticipatory Systems"* in order to describe this strange property of living systems:

> I was amazed by the amount of anticipatory behaviour observed at all levels of the organization of living systems (...) systems that behave as true anticipatory systems, systems in which the present state changes according to future states, violate the law of classical causality according to which changes depend solely on past or present causes. We try to explain this behaviour with theories and models that exclude any possibility of anticipation. Without exception, all the theories and biological models are classical in the sense that they only seek causes in the past or present.[14]

The neurologist Antonio Damasio discovered that neural damages localized in the prefrontal regions of the brain, especially in the ventral and medial sectors and in the right parietal region, are systematically associated with decision-making deficits. These damages are linked with the impaired perception of emotions and feelings. It seems that without feelings, the process of reasoning and decision making is no longer oriented towards the future.

In 2007, as part of her PhD research, Antonella Vannini formulated the following testable hypothesis: *If life is supported by syntropy, the parameters of the vital systems that support life, such as the autonomic nervous system, should show retrocausal activations.*

Various experiments had already shown anticipatory pre-stimuli reactions of skin conductance and heart rate. One of the first studies was performed in 1997 by Dean Radin,[15] who monitored heart rate, skin conductance and fingertip blood volume in subjects who were shown a blank screen for five seconds and a randomly selected calm or emotional picture for the following three seconds.[5] Radin found significant differences in the autonomic parameters preceding the exposure to emotional versus calm pictures. In 2003 Spottiswoode and May replicated Radin's experiments, adding controls to exclude artifacts and alternative explanations. Results showed an increase in skin conductance 2–3 seconds before emotional stimuli were presented ($p=0.0005$).

Similar results have been obtained by other authors, using various parameters of the autonomic nervous system, for example: McCraty, Atkinson and Bradley,[16] Radin and Schlitz,[17] and May, Paulinyi and Vassy.[18]

Antonella Vannini conducted four experiments using heart rate measurements in order to study her retrocausal hypothesis of the parameters of the autonomic nervous system. Results were strong both from a quantitative and a statistical point of view. Detailed information is available in the book *Retrocausality: Experiments and Theory.*[19]

It is interesting to note that although the anticipatory reactions of the parameters of the autonomic nervous system are strong and clear, this information is processed at an unconscious level and is not transferred to the conscious level of the rational brain. The dissociation between feeling the future and the ability to use this information rationally was assessed.

The Mind

In his paper "Chaos, Quantum-transactions and Consciousness"[20] Chris King starts from Einstein's energy/momentum/mass equation and speculates that free will arises from the fact that we are faced with bifurcations between information arriving from the past (causality-entropy) and in-formation arriving from the future (retrocausality-syntropy).

Retrocausality ($-E$) would be felt, whereas causality ($+E$) would be perceived and processed rationally. This constant antagonism between feelings and rationality forces one into a state of free will.

Supercausal model of free will

Free will is usually considered to be at the basis of all the actions of human beings, but it absolutely contradicts the assumption that only classical causality and determinism are real. Since the forward- and the backward-in-time energy solutions are perfectly balanced, similar amounts of information and in-formation are received. This might explain the perfect division of the brain into two hemispheres.

We can replace the previous drawing of the human being with that of the two hemispheres of the brain, where the left hemisphere is the seat of the "forward in time" rational brain and the right hemisphere is the seat of the "backward in time" intuitive brain.

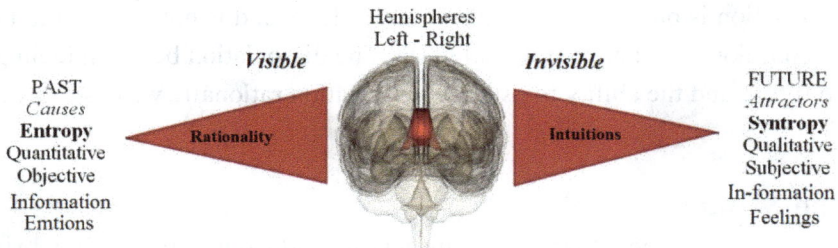

The rational brain is analytical and it is characterized by objective and quantitative information, whereas the intuitive brain is global and it is characterized by subjective and qualitative feelings.

The Syntropic model of consciousness adds to the brain, the autonomic nervous system and the attractor, and describes the mind as organized on three levels:

- the *conscious mind* which is associated with the brain;
- the *unconscious mind* which is associated with the autonomic nervous system;
- the *super-conscious mind* which is associated with the attractor.

The *conscious mind* to which we are tuned during the time we are awake, connects us to the physical reality. The conscious mind chooses between feelings that come from the autonomic nervous system and in-formation that comes from the physical plane of reality. This continuous state of choice is at the basis of free will.

Conscious
Free will

Unconscious
Automated processes

Superconscious
Intuition
Finality / Meaning

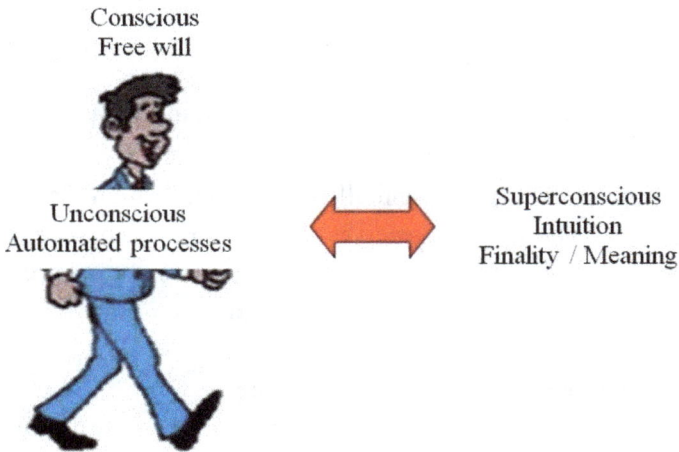

The *unconscious mind* governs the vital functions of the body, therefore called involuntary, such as heartbeat, digestion, regenerative functions, growth, development and reproduction. In addition, it implements highly automated programs that allow us to perform many complex tasks, without having to think continuously about them, such as walking, riding a bicycle, driving, etc. The autonomic nervous system supplies the body with syntropy and is therefore the seat of feelings that inform us about the connection with the attractor. The unconscious mind can be accessed during dreams, or using techniques of relaxation and altered states of consciousness such as hypnotic trance.

The *superconscious mind* is our attractor, the source of syntropy, the energy of life, which guides us towards wellbeing and happiness. The superconscious mind provides us with a mission, a purpose, and uses intuitions, insights, dreams and visions. It provides intelligence, knowledge and answers to problems. It leads towards more intelligent and perfect designs, which are the outcome of the contribution of all the individuals who share the same attractor.

Attractors

In 1963 the meteorologist Edward Lorenz[21] discovered that small changes in the initial conditions can amplify and make any prediction impossible. Lorenz also found the existence of an attractor which greatly amplifies the initial perturbations. He described this situation with the words: "*a flutter of a butterfly's wings in the Amazon can cause a hurricane in the United States.*"

Attractors generate fields, such as the gravitational field, the magnetic field, the electrical field, or the quantum field. These fields are invisible, but bring parts together and allow a communication that is invisible and intelligent. The attractor receives all our information/experiences and naturally selects only those that are advantageous and disseminates them to all related individuals. This mechanism has been described by Rupert Sheldrake as "morphic fields."

The evolutionary paleontologist Teilhard de Chardin discovered that species evolve within tracks informed by attractors. He formulated the hypothesis that life and the entire universe evolve towards a unifying point, an attractor, which he named the "Omega Point." He saw the need to extend science to an anti-entropic energy with syntropic properties:

> Reduced to its essence the problem of life can be expressed like this: accepting the two principles of conservation of energy and entropy, how can they assimilate without contradiction, a third universal law (which is expressed by biology), that of the organization of energy?...the situation becomes clear when we consider, at the basis of cosmology, the existence of a sort of anti-entropy... In other words, not just one kind of energy, but two different energies; two energies which cannot transform directly one into the other, because they operate at different levels... The behaviour of these two energies are so completely different and their manifestations so completely irreducible that we might believe they belong to two completely independent ways of explaining the world. And yet, as the one and the other, are in the same universe, and evolve at the same time, there must be a secret relationship.[22]

Experiments show that water is the substance through which the attractor of life manifests. This might help to explain how homeopathic remedies work. Homeopathy was discovered by Samuel Hahnemann, a German physician, and it is based on the so-called principle of similarity, according to which the appropriate remedy for a particular disease is given by a substance that, in a healthy person, causes symptoms similar to those seen in the ill person. Homeopathy uses water for its remedies. The patient is administered a remedy in which the substance (or active ingredient) is strongly diluted in water: the higher the dilution, the greater the power of the remedy.[23]

The paradox is that the most powerful homeopathic remedies are those that cannot contain a single molecule of the active ingredient. Having removed the active ingredient by dilution, it is believed that the effects are due to a placebo effect and not to an actual effect of the remedy. But attractors work in the opposite way! The active ingredient, when in water, follows the butterfly effect and the most diluted ones, when correctly placed in the attractor become the most powerful ones.

Western medicine refuses homeopathy since the effects cannot be explained in a classical way, but still the results are tangible and can be tested experimentally.

Love

We continuously produce maps of the physical world which entropy has inflated towards the infinite. Syntropy, on the contrary, has focused consciousness towards the infinitely small. When we compare ourselves to the physical world we find to be equal to zero and this is incompatible with our feeling of being alive:

$$\frac{I}{Universe} = 0$$

Compared to the universe, I am equal to zero: Hamlet's "*to be or not to be.*" Being zero is incompatible with our feeling of being alive. This leads to feeling worthless, aimless and depressed. We need to provide a purpose to our life, otherwise we go nowhere. Many think they can do this by increasing the numerator:

$$\frac{(I+judgment+wealth+popularity+power+\cdots)}{Universe} = 0$$

We can increase the numerator indefinitely, but compared with the infinity of the physical universe the result is always zero, and we continue to feel depressed and meaningless.

The solution is provided by the Theorem of Love:

$$\frac{I \ x \ \cancel{Universe}}{\cancel{Universe}} = I$$

When I compare myself to the universe and I am united with it through love, I am always equal to myself

The Theorem of Love tells that:

- only when our inner world unites with the outer world can we overcome the identity crisis;
- love provides this unity (I x Universe);
- love allows us to shift from duality (I = 0) to non-duality (I = I).

Love is synonymous with unity. When we converge our heart fills with warmth, joy and love. But when we diverge we experience void, pain and the conflict between being and not being. Love provides the aim and meaning to life.

Love is the attractor of life that brings parts together. The unity of our Self is strengthened when we love, when we are converging. When, on the contrary, cohesion diminishes, our personality shatters. Love is therapeutic since it brings together our parts and makes them cooperate.

It is interesting to note that since love reinforces the Self, it also increases individualization and differentiation; nonetheless it leads towards unity. It seems a contradiction, but unity and diversity go together.

Epilogue

The scientific revolution that was started by Newton and Galileo divided culture in two parts: on the one side science, capable of studying the entropic aspects of reality, and on the other side religion, dedicated to the syntropic aspects of reality, such as the soul and the final causes.

Extending science to syntropy implies a profound change in the cultural balance between science and religion, which Fantappiè describes as follows:

> Let us conclude by looking at what we can say about life. What makes life different is the presence of syntropic qualities: finalities, goals, and attractors. Now as we consider causality the essence of the entropic world, it is natural to consider finality the essence of the syntropic world. It is therefore possible to say that the essence of life is the final causes, the attractors. Living means tending to attractors. But how are these attractors experienced in human life? When a man is attracted by money we say he loves money. The attraction towards a goal is felt as love. We now see that the fundamental law of life is this: the law of love. I am not trying to be sentimental; I am just describing results which have been logically deduced from premises which are sure. It is incredible and touching that, having arrived at this point, mathematical theorems start speaking to our heart!

Fantappiè stated that nowadays we see written in the book of nature—which Galileo said was in mathematical characters—the same laws of love that we find written in the holy books of the major religions.

> ...the law of life is not the law of hate, the law of force, or the law of mechanical causes; this is the law of non-life, the law of death, the law of entropy; the law which dominates life is the law of finalities, the law of cooperation towards goals which are always higher, and this is true also for the lowest forms of life. In humans this law takes the form of love, since for humans living means loving, and it is important to note that these scientific results can have great consequences at all levels, particularly on the social level, which is now so confused.... The law of life is therefore the law of love and differentiation. It does not move towards leveling and conforming, but towards higher forms of differentiation. Each living being, whether modest or famous, has its mission, its finalities, which, in the general economy of the universe, are important, great and beautiful.

Endnotes

1 Ulisse Di Corpo and Antonella Vannini: www.sintropia.it

2 Auffray J.P., *Dual origin of E=mc²*:http://arxiv.org/pdf/physics/0608289.pdf

3 Poincaré H,. *Arch. néerland.* sci. 2, 5, 252–278 (1900).

4 De Pretto O., *Lettere ed Arti*, LXIII, II, 439-500 (1904), Reale Istituto Veneto di Scienze.

5 Anderson C.D. (1932) *The apparent existence of easily deflectable positives*, Science, 76:238 (1932);

6 Fantappiè L. (1942) *Sull'interpretazione dei potenziali anticipati della meccanica ondulatoria e su un principio di finalità che ne discende*. Rend. Acc. D'Italia, n. 7, vol 4.

7 Van Flander T. (1996), Possible New Properties of Gravity, Astrophysics and Space Science 244:249–261.

8 Van Flander T. (1998), *The Speed of Gravity What the Experiments Say*, Physics Letters A 250:1–11.

9 Van Flandern T. and Vigier J.P. (1999), *The Speed of Gravity—Repeal of the Speed Limit*, Foundations of Physics 32:1031–1068.

10 Mario Ludovico, Syntropy: definition and use, *Syntropy Journal*, 2008, 1, (139–201)

11 Jung C.G. (1951), *Synchronicity—An Acausal Connecting Principle*, Princeton University Press, www.amazon.com/Synchronicity-Connecting-Principle-Collected-Bollingen/dp/0691150508

12 Wheeler J. e Feynman R. (1945) Interaction with the Absorber as the Mechanism of Radiation, *Review of Modern Physics* (17);

13 Damasio A.R. (1999), The Feeling of What Happens. *Body and Emotion in the Making of Consciousness*, Heinenann, London 1999;

14 Rosen R. (1985) *Anticipatory Systems*, Pergamon Press, USA 1985;

15 Radin D.I. (1997), Unconscious perception of future emotions: An experiment in presentiment, *Journal of Scientific Exploration*, 11(2): 163–180.

16 McCraty R (2004), Atkinson M and Bradely RT, Electrophysiological Evidence of Intuition: Part 1, *Journal of Alternative and Complementary Medicine*; 2004, 10(1): 133–143.

17 Radin DI (2005) and Schlitz MJ, Gut feelings, intuition, and emotions: An exploratory study, *Journal of Alternative and Complementary Medicine*, 2005, 11(4): 85–91.

18 May EC (2005), Paulinyi T and Vassy Z, Anomalous Anticipatory Skin Conductance Response to Acoustic Stimuli: Experimental Results and Speculation about a Mechanism, *The Journal of Alternative and Complementary Medicine*. August 2005, 11(4): 695–702.

19 Vannini A and Di Corpo U, *Retrocausality: experiments and theory*, ISBN 9781520275956.

20 King C.C. (2003) "Chaos, Quantum-transactions and Consciousness." *NeuroQuantology*, Vol. 1(1): 129-162;

21 Lorenz E. (1963) Deterministic Nonperiodic Flow, *Journal of the Atmospheric Sciences*, 1963, Vol.20, No.2, pp.130-140;

22 Teilhard de Chardin P. Le phénomène humain. Ed. du Seuil 1955.

23 Paolella M., Homeopathic Medicine and Syntropy, *Syntropy Journal*, 2014 (2): 1-29

Bioenergy Detection via fMRI

Wagner Alegretti

Introduction

Since ancient times, different areas of human knowledge and culture mention a form of subtle energy, parallel to the materiality of the physical day-to-day world, but not so far from human reality that it would not be felt and known by more sensitive people or those aware of its existence. It has received several names over the centuries, such as prana (Yoga), chi/qi (Chinese medicine), orgone (Reich), magnetic fluid (Mesmer), vital fluid (Kardec), life force (Hahnemann), astral light (Blavatsky), etheric force (dowsing), and more recently biofield or bioenergy, as it is called here.

This subtle energy has eluded modern science, not only with regard to its detection and measurement, but also with respect to its theoretical conceptualization and modeling, yet there have always been reports of people claiming to be able to feel or see this form of subtle energy in living things, the environment, and in other people. Sometimes people have also claimed to be able to apply it to improve their own quality of life or that of others around them, including in situations such as spiritual healing.

Studies have been carried out in various disciplines, including parapsychology, alternative medicine, and biology (with concepts like morphic resonance and biophotons) in order to clarify the nature of this energy. But in spite of millennia of demonstrated cases and of the scholarly research of pioneers like Wilhelm Reich, Semyon Kirlian, and Konstantin Korotkov, the reality of bioenergy has not been accepted by the scientific community.

This chapter will describe various experiments I have carried out since 1984 in order to detect and measure this bioenergy to better understand it and its role in the expression or manifestation of consciousness.

While studying electronic engineering in the university, I used various instruments and components available at the time, such as magnetometer, Geiger counter, semiconductors, and others, but without relevant or consistent results. Yet, based on the assumption that living things, including humans, are natural transducers of bioenergy, and since physical life would itself be the most substantial manifestation of such form of energy, and also knowing about the ability of many people throughout the centuries to feel it, I started with the hypothesis that the answer to my quest for a bioenergy detector lay in the exploration of organic matter.

To avoid the use of living things, due both to ethical concerns and the unreliability of the results due to interference of the natural dynamics of the internal metabolism and even to behavioral changes, organic substances *in vitro* were chosen. Proteins were selected as being universal in the central processes of life, and also because of their flexibility in reactions. Indeed, their behavior depends not only on their chemical composition, but also on their particular 3D shape or folding.

Transducer prototypes were built based on an aqueous gel of collagen, of which its electrical resistance was measured. Some promising results were obtained with this arrangement, but unfortunately, for various reasons this line of research had to put on hold. In 1990, I shared the results of my research done from 1984–1988 (Alegretti, 1990) during a conference entitled "Bioenergy Technology" at the 1st International Congress of Projectiology (Rio de Janeiro, Brazil). In that presentation, the principles of "bioenergy technology" were also presented along with the initial experimental results, discussions about its relevance and applications, and also the planning of the next phases for its development.

My subsequent investigations employed a resource widely exploited in medical analysis—the nuclear magnetic resonance imaging (MRI)—mainly a variation of it that allows the monitoring of the functioning of the brain during specific tasks: the functional magnetic resonance imaging (fMRI). The fMRI is based on the BOLD (Blood Oxygenation Level Dependent) technique to register the change of magnetic resonance of the hemoglobin when passing from the oxygenated state (arterial blood) to deoxygenated (venous blood). This technique has been used in many scientific studies of brain function and is also used in clinical exams, especially as a way to get previous data before certain brain surgeries.

I decided to employ the fMRI to monitor what happens in the brain when someone consciously and willfully controls his bioenergy to provoke an internal intensification of one's natural energetic vibration, a phenomenon known as vibrational state (VS) (Trivellato, 2008). The VS has the advantages not only of creating an unusual intensification of the personal bioenergy field, but also of being produced by the will through a standardized technique called VELO (voluntary energy longitudinal oscillation). This technique can be learned, developed, and applied by anyone (Trivellato, 2017). In 2008, I published a paper in the *Journal of Consciousness** (Alegretti, 2008) proposing this particular research and discussing its methodology, possible benefits, and applications. As explained below, this line of research was expanded to also understand what happens in the brain during the intentional absorption and transmission of bioenergy, including to different media such as fMRI phantoms, a potato, and an egg.

As bioenergy is claimed by many to be a central aspect of consciousness manifestation, not only through the expression of biological life itself, but also in a class of phenomena called paranormal, non-physical, extra-physical, psychic or parapsychic, its detection and evidencing could provide more solid evidence for the objective nature of consciousness, as understood in the so-called consciential paradigm (Vieira, 2002). According to this paradigm, the bioenergy is the bridge between the nonphysical consciousness (independent from the body) and the physical world of the ordinary forms of matter-energy (a form of "hard problem of consciousness").

The data and preliminary qualitative results from the three series of experimental sessions already conducted, presented below, seem to point towards the validity of this hypotheses and the viability of this research and methodology.

In parallel to the direct bioenergy detection line of research, some pilot experiments with electroencephalography (EEG) were also conducted to search for neural correlates of some consciential states and bioenergy procedures. In 1991, I had the first opportunity of doing some personal experiments with the lucid projection (out-of-body experience)

---

* Formerly known as: *Journal of Conscientiology*

while monitored by an EEG and other devices for measurement of several physiological parameters, in a sleep study laboratory of a hospital in the city of Porto Alegre, Brazil. On that occasion, I also established the VS voluntarily to allow for the observation of the changes in brain waves patterns that could be generated as a result of this process. That showed an unusual regimen of brain waves, also with a predominance on high frequency (gamma) waves.

In December 2007, another series of experimental sessions with EEG was conducted, with Nanci Trivellato and myself as subjects, at a neuroscience laboratory in the city of Natal, Rio Grande do Norte, Brazil. Those experiments focused on registering brain activity through digital EEG during the production of VSs and partial projections (Alegretti, 2008). Similar results were observed.

Continuing this specific line of investigation, a pilot study with more advanced EEG equipment and better scientific support has recently being initiated in partnership with the TransTech Lab at the Sofia University in Palo Alto, California, USA. Again, the occurrence of high frequency gamma waves was confirmed.

Another valuable source of information for this line of study has been the research done through the bioenergy evaluations of individuals, mainly of students of the course "Goal: Intrusionlesness," given by Nanci Trivellato and me since 2002 at the International Academy of Consciousness (IAC). Those individual evaluations produced observations, findings, and further questions about the VS and bioenergy procedures, generating new hypotheses and allowing for the perfecting of the experimental protocol of the research described in this paper (Trivellato & Alegretti, 2005).

Some of the preliminary results and analyses of this original fMRI research have already been shared with fellow researchers such as Dean Radin (Institute of Noetic Sciences), Beverly Rubik and Harry Jabs (Institute for Frontier Science), Ivan Lima (North Dakota University) and members of the Foundation for Mind-Being Research, in Palo Alto, California, USA.

The EEG line of research recently inspired Rute Pinheiro (Universidade Federal do Rio Grande do Norte, Brazil) to initiate and conduct a similar investigation (Pinheiro, 2013).

Hypotheses

Assuming the Consciential Paradigm—that consciousness is objective, multidimensional, another property of the universe, neither reducible to matter-energy, nor a mere product or epiphenomenon of it—this research is founded on three basic hypotheses:

1. Bioenergy is real and objective, being able to interact with matter and other forms of energy, as in the manifestation of Life;
2. The VS, and other bioenergy regimens or manifestations, are objective conditions, not just imagination, illusion, or sensory hallucination of the practitioner;
3. The VS and other bioenergy procedures and regimens are, in certain circumstances, accompanied by detectable changes in the human brain or can cause alterations in such (some temporary, and others perhaps more permanent).

Thus, this research was initiated to study the VSs and bioenergy procedures in which effects can directly reach the physical body, while recognizing that the VS possibly happens primarily at the level of the subtle energy system, which would act as an interface between the (non-physical) consciousness and the physical body.

Objectives of the Research

The experimental development of studies in this area has sufficient merit, since it would allow for promoting the following possible relevant results and discoveries, among others. So, within a broad view, in short, mid and long terms, the objectives of this line of research are:

- Create means for the detection of bioenergy;
- Measure such bioenergy;
- Gather a sufficient amount of different and converging evidence that support the theory of the objectivity of the bioenergy (see hypothesis 1);

- Understand the biological, and more specifically, neurological effects of bioenergetic processes like the VS and bioenergy in general;
- Identification, categorization and description of the neurological effects provoked by the bioenergy procedures or concomitant to it (see hypothesis 3);
- Demonstration of the VS and other bioenergy regimens as real and objective energetic phenomena (see hypothesis 2);
- Characterize the VS as a stand-alone state, different from other neurological, physiological or mental states;
- Collection of data and findings for a better comprehension of the VS itself;
- Better understanding of the processes and factors involved in the development and effective bioenergy procedures and techniques, allowing for the generation of more efficient didactical methods and more exact descriptions, capable of promoting better energetic self-control for the community of practitioners;
- Classification of the VS and other bioenergy regimens according to the level of effect on the body, and then according to their types and the intensity of the effects;
- Clarify the mechanisms of the consciousness-physical body interface, to promote understanding of how consciousness (a non-physical entity) interacts with the physical dimension, or, in other words, how it can control and sense the physical body and universe;
- Search for and develop new therapeutic applications of the VS in particular, and bioenergy in general;
- Build a bridge between the study of consciousness under a multidimensional paradigm and the more physical neuroscientific approaches, thus enriching both;
- Investigate bioenergy's characteristics and properties and establish the general laws that govern it (analogy: history of the study of electricity and magnetism up to the Maxwell equations) to create a theoretical framework capable of predicting findings for understanding of bioenergy within a multidimensional view of the nature of consciousness.

fMRI Experiments

The physical repercussions of the VS and other bioenergy procedures and regimens can in principle be studied according to different criteria and techniques. The ideal would be through a bioenergy technology that would allow for the direct detection and measurement of the energies, but this technology does not yet exist. As an indirect method, however, the fMRI appeared to be a good choice for initial tests due to its advanced design and the fact of its being based on a pure quantum effect (the alignment of nuclear spins in a magnetic field), and its allowing monitoring of changes in brain functioning,

In other words, to allow for the realization of a more objective examination of the effects of the VS and a systematic comparison of the results, the most effective and consistent method seemed to be the detection of the neurophysiological alterations of the practitioner through the best technology for analysis of neurological functions available today. Another important point is that the method allows for the replicability of the experiments by any researcher, even those with a more skeptical approach or those who have never felt or produced bioenergy procedures or effects.

fMRI experiments were performed in 3 series, as listed below, using MRI systems graciously lent by generous clinics and radiologists in Brazil, who allowed their use over weekends and during other free periods of time when the systems were not being used to perform clinical exams. Also, the chief radiologist of each clinic took part in most or some of the experiments of each series, monitoring the results and contributing suggestions and advice. Furthermore, MRI technicians were made available for the running of the tests.

- Series 1: 2009–December—1T Philips machine
- Series 2: 2010–March—1T Philips machine
- Series 3: 2014–December—3T Philips machine

As described above, these experiments started with the initial objective of studying neurological changes of the brain during the procedure of creating the VS and after it. Nonetheless, as will be explained later, it broadened to also encompass the effects of bioenergy over matter and the mechanism of consciousness-matter interaction via bioenergy thus far observed on water solutions, potato, and egg.

Basic Protocol

Each individual experiment (data acquisition session of the fMRI system) was planned to be divided into 3 periods, as follows:

- 1st period: initial resting (inaction) with the subject or "energizer" being conscious but as relaxed as possible, to establish a baseline or reference state;
- 2nd period: action (this was different for each group of experiments or sessions, as described below);
- 3rd period: post-resting (inaction), with the subject or "energizer" again being conscious but as relaxed as possible, to establish an "afterglow" or a second reference state, but also to test any possible lingering or delayed effect of the previous action period.

The average duration of the individual experiments was of 3 minutes, divided equally in 3 periods of 1 minute.

As it is the case in all fMRI tests, at the beginning of each series of experiments with a subject, be it a person or object, a regular MRI test for static, anatomical data acquisition was run, which took between 40 to 50 minutes. Later, the fMRI (BOLD) tests were run and the images imposed over the static or "anatomical" images.

In the cases when people were the subjects, after the static MRI imaging, a standard "finger tapping" test was run to verify the proper functioning of the system, adjusting the parameters to avoid a too low sensitivity (that would cause no images during the action phase) or a too high sensitivity (that would reveal noise or artifact images during the inaction periods).

As in many experiments of this type, an intercom between the experimental (MRI) room and the control room was used to guide the subject's actions and to monitor the results. This allowed the subject to announce to the researchers beforehand what he wants to do, what according to him just happened, or what is taking place at each period, among other possibilities. It also allowed the MRI technicians to guide the experiment telling the subject the exact time to initiate or stop each of the 3 periods indicated above.

The recording and analysis of the brain activity in certain stages described below (apparently disconnected from the objective of the bioenergy experiments) have the purpose of serving as a 'control' reference for the analyses, since they allow comparison of the results with the data obtained from the voluntary and correct application of the VELO and possible implementation of the VS with great intensity.

In phase 1 of the experimental protocols, during the "action period" the following actions were performed:

1. Relaxation/rest only (resulting in continuous inaction during the 3 periods);

2. Conscious and voluntary rhythmic inhaling and exhaling only (as some people do while trying, erroneously, to perform VELO)—slow in the beginning and then with its gradual acceleration;

3. Conscious and rhythmic moving of the eyes up and down, with the eyelids closed (as some people do while trying, erroneously, to perform VELO)—slow in the beginning and then with its gradual acceleration;

4. Visualizing or imagining the rhythmic movement of bioenergy up and down, slow in the beginning and then with its gradual acceleration, simulating VELO, without real intent or action to move energy;

5. Scanning of the perception of the sensations of the body, up and down. In other words, only sweeping of attention and perceptive focus through the body—slow in the beginning and then with its gradual acceleration. In this case, the subject tries to concentrate exclusively on the part of the body that is being focused on. Such focus moves smoothly, continually and cyclically up and down along the body (from feet to head and vice-versa);

6. Effective initiation of the VS through the correct and vigorous application of the VELO.

In accordance with the above discussion, this protocol can be used with any data collection resource, like fMRI or EEG, or even, in the future, with a direct bioenergy detection system.

Even though the procedure described on Stage 1 is unnecessary for the production of the VS (though not counterproductive), its recording and study are essential to establish the baseline; that is, the basic resting neurological condition specific and particular to each participant. This resting condition will be an important reference for posterior comparisons and analyses.

The six experimental steps here described were planned to make possible, in the data analysis, the basic strategy of subtracting from the data set relative to Stage 6 the signals obtained in the previous stages. In this way, it becomes possible to determine the profile of the VS itself by separating its neurologic 'signature' from the other associated neurological processes, be they natural or derived from the technique application (assuming here that the neural structures supporting cognitive and behavioral processes combine according to a simple additive logic).

The steps described in the Stages 2 to 5 aim at simulating a pseudo-execution or erroneous execution of the technique of installation of the VS, having been included to take into consideration the habits (some of them inappropriate) common in the application of the VELO technique. Evidently, the data obtained during those stages have the objective of being more than just 'noise to be removed,' since the careful analyses of those can lead us to a better understanding of the mechanisms of the technique of the VELO and of the pro and con factors in the attainment of the VS. Furthermore, they allow for a clearer verification of the potential influence of those physical (body) or mental 'crutches' over the execution of the VELO and the VS.

Results

With VELO and VS:

- For actions 2 through 5 listed above there was no relevant difference when compared to the "action" 1 (whole session of resting) or the inaction periods before and after each of those respective actions (there was very little or no BOLD activation).

- VELO produced significant BOLD activation, and when VS occurred the activation was comparatively stronger.
- During VELO, and even more during VS, there was intense activation of many different areas of the brain (see Figs. 1 and 2), distinct from and perhaps even stronger than that of normal actions or tasks, but certainly stronger when compared to the previous finger tapping (see Fig. 3).
- For the VELO experiments, no specificity or pattern of active areas of the brain or cerebellum were observed for the same subject, or even across the three subjects tested.
- During VELO and VS, images (BOLD-like signal) appeared outside of the skull region, something that, in principle, should not happen. The first explanation was obviously artifacts. Nonetheless, these extra-cranial signals remained consistent throughout most of the VELO-VS sessions during series 1 and 2, but not during series 3. It is important to remember that the coil room (exam room, where the subject stays during the tests) is heavily shielded to avoid interferences, inwards or outwards. Even the possibility of a

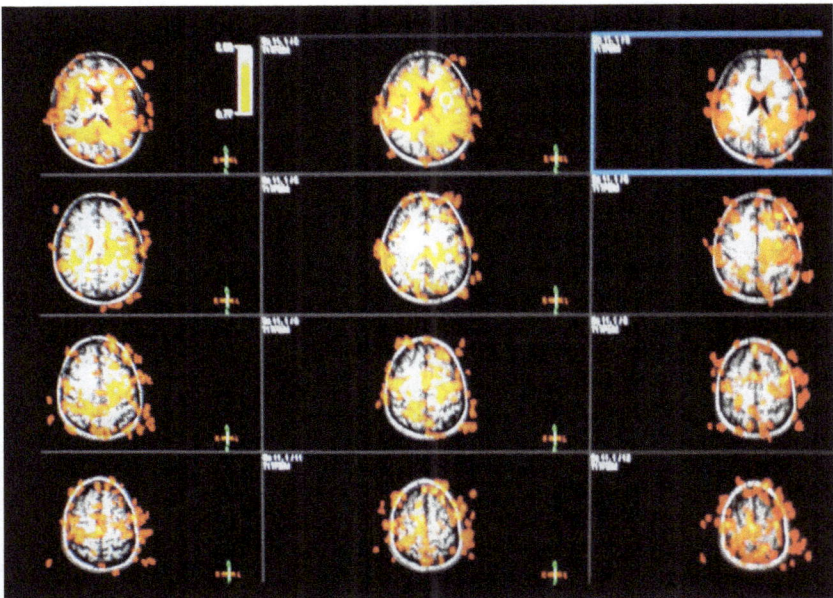

Figure 1 – fMRI images of the brain with VS, strong and generalized BOLD.
Series 1: 1T machine

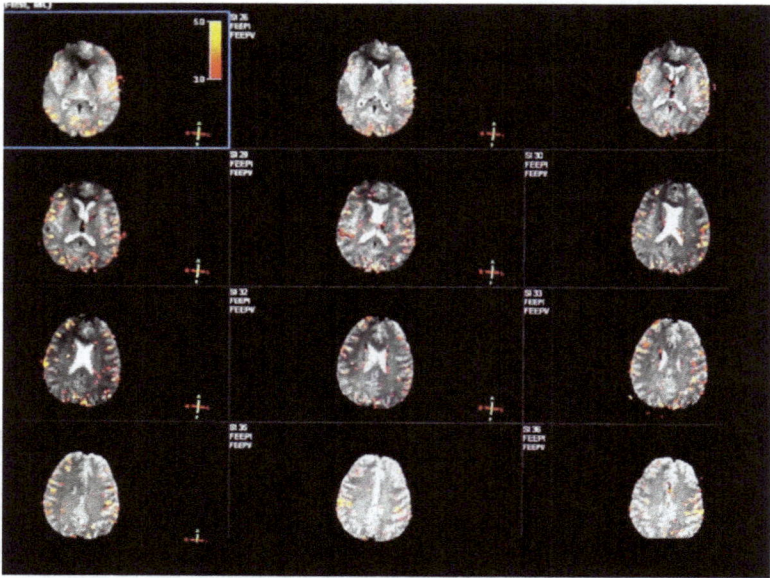

Figure 2 – fMRI images of the brain when doing VELO, spread BOLD signals.
Series 3: 3T machine

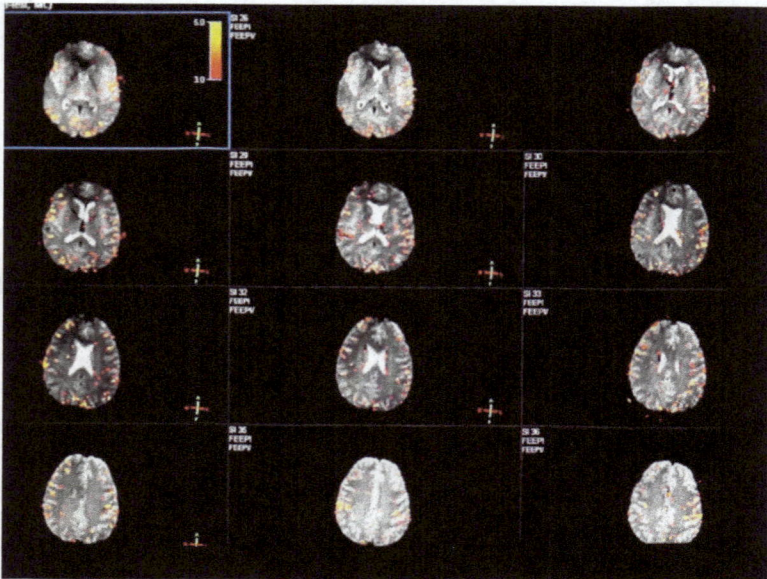

Figure 3 – fMRI images of the brain when doing finger tapping, a reference;
Series 3: 3T machine

shower of cosmic rays was considered, but it was discarded when the same effect was repeated many times.

- It was considered that small head movements could have provoked the out-of-the-head images. During the series 1 and 2 the radiologist said he reviewed the data and confirmed there was no movement. Even so, extra experiments were conducted to try to control for that, and during the action period the subject produced small amplitude and fast trembling of the head by contracting strongly the neck muscles, even more pronounced than any micro movements possibly produced during normal VELO and VS. A few and faint BOLD images were produced out of the head, but significantly weaker than with the real VELO and VS.

- Due to the unexpected finding of the out-of-the-head signals after the first series of experiments, it was decided to expand the original plan of studying only the VELO procedure and VS, by having the subject actively irradiate bioenergy from the head. The result was the production of even stronger out-of-the-head images, across all experiments of this sort (Fig. 4).

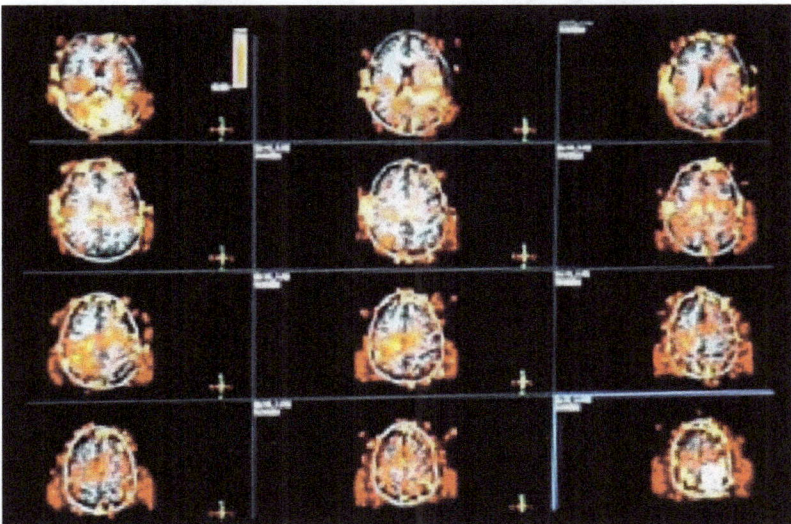

Figure 4 – fMRI images of the brain when doing irradiation of energy from the head; Series 1: 1T machine

- To explore further the possibly important finding of an apparent direct effect of bioenergy over a non-organic (and non-living) medium, as the active subject (the "energizer") for this particular series of experiments, I decided to use a MRI phantom as the "the fMRI or passive subject" inside the secondary coil (the one normally put around the head of the person when under MRI examination). The phantom used during the series 1 and 2 (Fig. 5) was basically a plastic bottle with water containing copper sulfate, sodium chloride, sulfuric acid, and *arquad* (a surfactant and preservative agent).

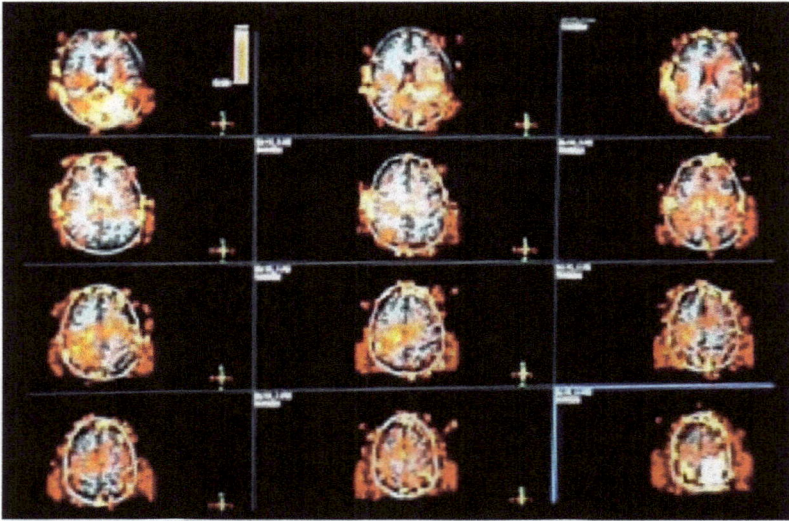

Figure 5 – Bottle used as fMRI phantom for the transmission experiment (showing composition). Series 2: 1T machine

- With the phantom inside of the coil and the energizer lying on the bed with abdomen down, arms stretched above the head with both hands positioned out and at the sides of the secondary head coil, further experiments were run (Fig. 6).
- The hands were kept out to avoid or at least reduce any possible influence of hand movements or blood circulation changes during the "action period" of the experiment. A strong BOLD signal occurred again during the energization period, which stayed and got

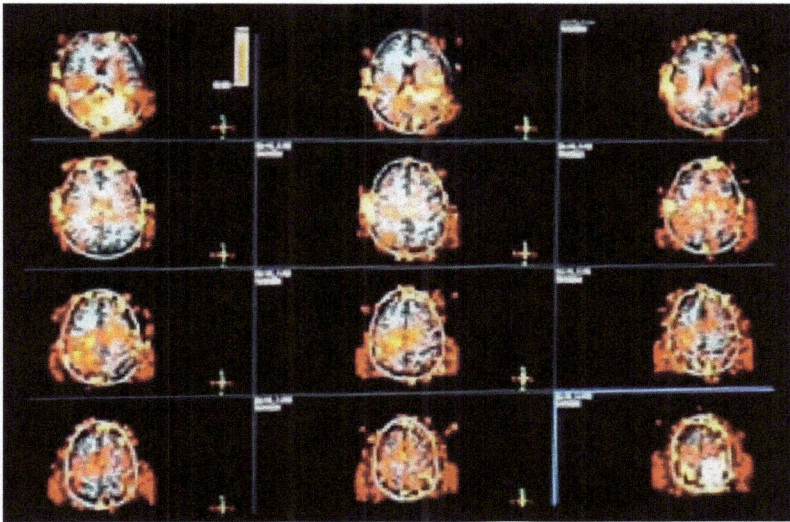

Figure 6 – Phantom-energizer-machine setup for the transmission to the phantom;
Series 2; 1T machine

slightly stronger even during the 3rd period (Fig. 7). This result
was confirmed with all of the other repetitions (sessions) of the
same type of experiment. During the sessions with no externalization of bioenergy during the "action period," but with exactly
the same setup, (with hands still at the external side of the secondary coil surrounding the phantom) there was no BOLD signal. So, the mere presence of the hands or their micro-movements
were not producing BOLD signal and corresponding images.
During the 3rd series of experiments (3T machine) a different
phantom was used. It was spherical and the solution was of water
and chromium chloride. Surprisingly and unexpectedly there was
no relevant result. This could be because the head-neck coil was
being used (see the discussion in the next section) or due to the
specific chemical profile of it.

- During the 3rd series of experiments a potato was also used. No
  relevant results were obtained, just a few very little groups of pixels of BOLD image (Fig. 8).
- During the 3rd series of experiments an unfertilized chicken egg
  (a common type from a supermarket) was also used. The results

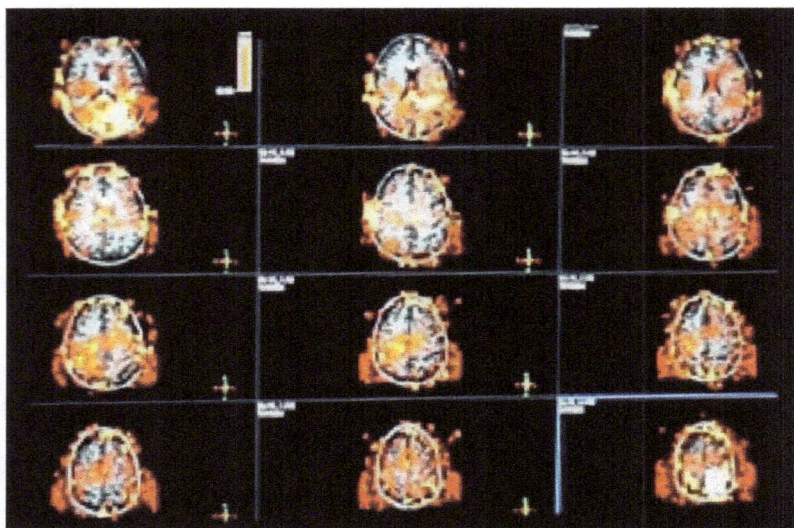

Figure 7 – fMRI images of the phantom with transmission of bioenergy from the energizer. Series 2; 1T machine

obtained were the most significant for being very intense and very synchronic with the on and off periods of the experiments. To make the results even more reliable, in this case the hands of the energizer were kept not at the sides of the head coil, but further down along the bed, about 10 cm far from the base of the coil. During the initial inaction-resting period there was no BOLD activation, but 4 seconds after the beginning of the energization (transmission of bioenergy to the egg) the BOLD signal started to appear and kept intensifying up to the moment the energizer received the instruction to stop. Then, after about 5 seconds the signal started to fade until it disappeared. During the activation, the BOLD image revealed an internal structure inside of the egg (not noticed before), mainly inside of the yolk. See Fig. 9 (before) and Fig. 10 (during).

• Successive experiments with the same egg and energization procedure started to show a growing tendency for lingering effects of the bioenergy. In other words, with each experiment of this type there was a stronger and more lasting BOLD activation after the cessation of the exteriorization of bioenergy.

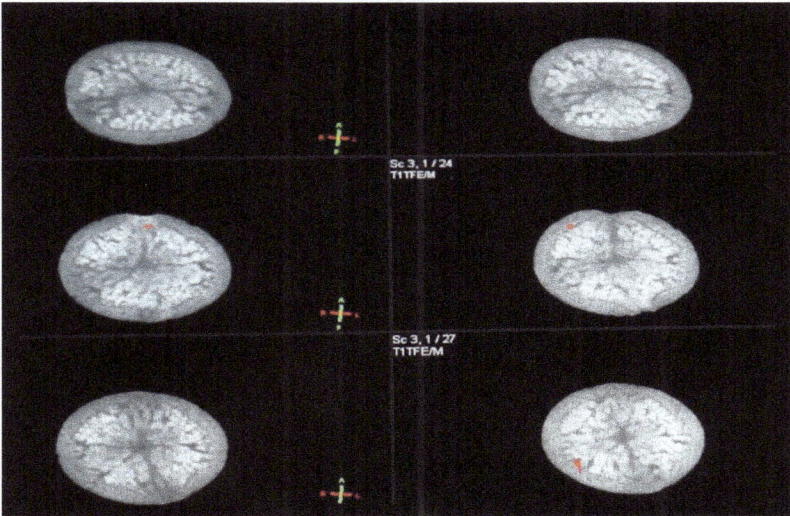

Figure 8 – fMRI images (detail) of the potato with transmission of bioenergy from the energizer; Series 3: 3T machine

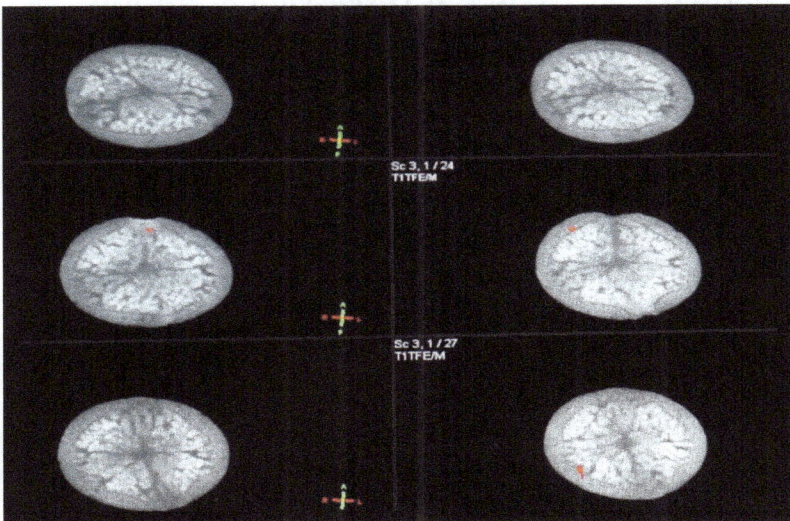

Figure 9 – fMRI images of the egg before transmission of bioenergy from the energizer (no BOLD signal present), the same after. Series 3: 3T machine

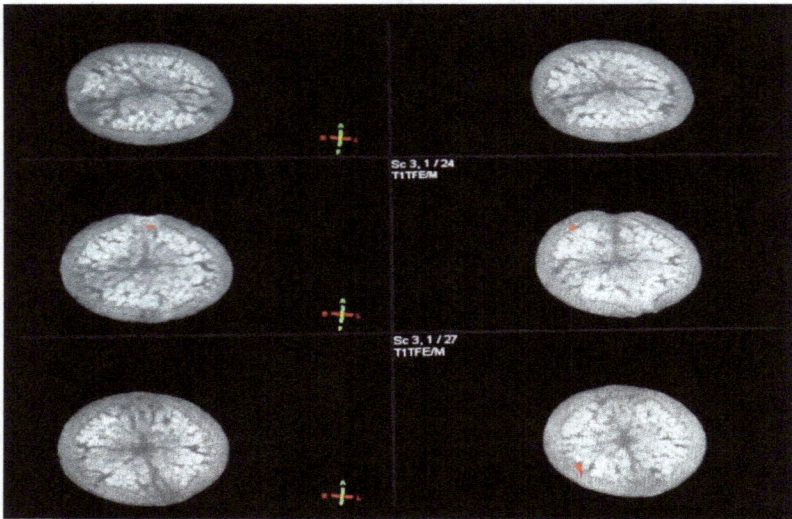

Figure 10 – fMRI images of the egg during transmission of bioenergy from the energizer (strong BOLD signal). Series 3: 3T machine

Preliminary Analysis and Discussion of Results

- The MRI head coil is more sensitive than the head-neck coil and should be the one used in this kind of experiment. The sessions performed with a head-neck coil, used only at the beginning of the 3rd series of experiments, with the 3T system, produced no relevant or reliable results. All experiments done with that coil had to be repeated with the more sensitive head coil. In spite of knowing that the head-neck coil was somewhat less sensitive, it was used to try to observe a wider area of the body and around of the head.

- With respect to the out-of-the skull images during VELO and externalization, could the intensified bioenergy field around the head be altering the MR (magnetic resonance) properties of one or more of the gases of the air (related to their spins or collective behavior of its molecules), to the point of producing a BOLD-like signal that can be picked by a machine able to detect magnetic variations, in a way similar to what happens with the hemoglobin? If so, which gas(es)?

- The same effect occurred with the phantom. Was the bioenergy acting equally on all the different types of atoms or molecules of the constituents of the phantom or only on one or a few of them? Which ones would be more responsive to the bioenergy? If such, what kind of "artifact" could be produced that way?

- In the case of the chromium chloride phantom (3rd series of experiments) the absence of BOLD activation could be because the head-neck coil was being used or due to the specific chemical profile of it. Also, if the activation seen in the other experiments was just an artifact (caused for instance by hand movements), why it did not happen in this case? Would this happen if I had used the more sensitive head coil?

- It is of course necessary to repeat this kind of experiment with many different types of MRI phantoms, preferably always using a coil of same sensitivity, to look for possible patterns.

- Why there was such a small result with the potato? There are water and many complex molecules in it. Why did none of them react significantly to the bioenergy that particular time? Again, if the activation seen in the other experiments were just artifacts (caused for instance by hand movements), why it did not happen with the potato?

- Why there was such a strong effect with the egg? Why, with the egg, unlike the phantom bottle, was there no lingering activation after cessation of the energization? Was it because of the medium per se, or could it be due to a particular difference in the data acquisition, analysis, and presentation by the fMRI system? What is the physical mechanism (detectable via the fMRI technique) that could explain these different observations? In that respect, it is interesting to observe that the spin of subatomic entities, which is the origin of the changes in magnetization observed via the fMRI technique, is a quantum observable having no classical analogue. Could it be then that the spin, with its typical non-spatial properties (Aerts & Sassoli de Bianchi, 2015), could provide a sort of bridge between ordinary matter-energy and the subtler bioenergy?

- As for the BOLD image revealing an internal structure inside of the egg, mainly inside of the yolk: which substances predominate there, that can be more strongly activated by bioenergy? Perhaps some specific proteins could be used in further experiments, to provide a more efficacious transducer (as discussed earlier), without the need for the complex and expensive fMRI in respect to the detection and measurement of bioenergy.

- If the lingering effect of the bioenergy is confirmed, would it change with different materials? Will it be possible to identify different bioenergy decay rates, for different materials?

- The occurrence of activation (BOLD images) in the air, bottle phantom and egg bring the very important question of whether the images detected in the brain during VELO and other bioenergy procedures are the result of the mind-brain willful action of moving the bioenergy, or if the brain is being affected by the bioenergy (or a combination of both things)? Can the brain (a physical organ) act on the energy of the possible subtler bodies of the consciousness, or is it the opposite (or even a combination of both)?

- As seen with these preliminary experiments, they mix neurology, biology, and physics in a way in which it is difficult to separate their respective domains of study. It is necessary to design experiments to isolate them as much as possible.

- Under the very strong field of the 3T machine it was significantly more difficult for the subject (myself) to "work" with or to move his bioenergy. Why? Was this due to the effect of a magnetic field 3 times stronger that with the system used during series 1 and 2? Or, could it be due to other yet to be identified physical factors, or non-physical ones, like the "vibes" (information-energy matrix left by previous patients) of that particular machine?

- It is worth mentioning that the limitation of time has been always an important constraint (for all the 3 different series) as the time available was not enough to run as many sessions of the same type as would be ideal, or changing subjects for all of them.

- These experiments are still somewhat exploratory. There is so far no theoretical framework with specific predictions to be tested. This certainly does not invalidate the experiments and their relevance. Many important scientific discoveries and progresses have happened this way, as with the famous Maxwell equations that describe electricity and magnetism in a unified way, which could only be derived following a series of exploratory experiments done by Volta, Galvani, Faraday, Marconi, Hertz, and many others. Quantum mechanics is another example of a theory that was created to deal with otherwise unexplained phenomena like the blackbody radiation, the photoelectric effect, and the electromagnetic radiation absorbed and emitted by atoms. And, of course, physics is always facing new explanatory gaps, requiring new theories, as is the case for instance today with the unexpected experimental observation of the so-called dark matter and dark energy.

Possible Future Experimental Steps

Based on the results obtained so far and also on the experience accumulated, I plan to conduct the following steps, in the short, mid, and long terms:

- Comparison with EEG experiments, under similar conditions (as described earlier, a pilot study is already on the way).
- During the next series of fMRI experiments, to focus more immediately on effects of bioenergy on materials, not the brain, for initial simplicity and better control of the experimental parameters, to gather knowledge for better understanding later, about what happens in the brain, discriminating what is cause and what is effect.
- Perfect protocol and technical conditions of the experiments, incorporating suggestions and criticism collected after publication of these preliminary findings and analyses.
- Test different materials to identify those most responsive to bioenergy. Based on the egg experiment, it could start with pure proteins like albumin, collagen or laminin.
- Use a 9T fMRI system (9 times stronger than the machine used in series 1 and 2, and 3 times the one used in series 3), to evaluate the effects of a more intense magnetic field over all the previous preliminary findings. For example: would it be even harder for the subject-energizer to move energies inside of such a machine? More powerful systems are more sensitive to small effects, but also more prone to noise and artifacts.

- Develop a protocol to measure intensity of the bioenergy based on the BOLD technique: standardized conditions, identification of an appropriate scale, quantitative analysis of the fMRI image, etc.
- Test effects of: time decay of the bioenergy activation; bioenergy accumulation; distance between energizer and medium; different materials put between the energizer and the medium (bioenergy shielding).
- Expand the fMRI and similar experiments to a wider and more diverse group of subjects or practitioners, to find common characteristics and determine average values.
- Deepening the research of other energetic maneuvers.
- Third-person experiments.
- Research of out-of-body experiences.
- Broadening of the research via the study of the possible spontaneous occurrence of the bioenergy changes in animals and the effects of bioenergy transmission to them.
- Comparative analyses (cross-analysis) of the fMRI study with the results of other similar researches (with the same objective, however, although with different methodology), such as for instance, surveys, individual interviews, and bioenergy evaluation of the subjects by sensitive and experienced bioenergy workers (as the research conducted during IAC's course "Goal: Intrusionlessness," already given to many groups) (Trivellato & Alegretti, 2005).
- Experiments with the protein transducer; Reich's orgone based instruments; gas discharge devices (Korotkov devices), for comparison.
- Neurological analyses of the VS and bioenergy procedures through the use of other technologies, like, for instance:
  - PET scan (Positron Emission Tomography)
  - fNIRS (functional Near Infrared Spectroscopy)
- Analyses of other effects of the VS and bioenergy (endogenous or exogenous) on the body, such as:
  - biochemical changes: hormonal and metabolic
  - influence over the immune response
  - epigenetic changes, or the pattern of expression of genes

(which ones would be activated; which ones would be deactivated; mechanisms of these changes)

- Development of a portable and simpler NMR device specific for bioenergy detection, measurement, and analysis, as the size, price and complexity of the medical MRI system is a direct result of the need of a big opening in the main magnet for a person to be put in and also for the generation of 3D image for clinical diagnostics.
- Development of a Bioenergy Technology, analogous to the development of the mechanical, thermodynamic, electrical-magnetic, and information technologies.
- Test the hypothesis of ectoplasm as being an exotic state of matter, as it seems to be created by bioenergy action.

Conclusion

The preliminary results so far are encouraging and point to evidence for the objective reality of bioenergies and their role in the consciousness' action over the brain and other forms of matter. They should be seen as meriting further experiments by other researchers looking for sustaining or falsifying these findings and their interpretation. It is very important to continue and progress with further experiments.

It is essential to emphasize that these results cannot yet be taken as definitive proof (if such a thing exists) of the objectivity of bioenergy and its manifestations. It is necessary that several other open-minded researchers replicate these experiments. If they reach the same or similar results, then a greater acceptance of these findings would be achieved.

As the fMRI is very prone to noise, artifacts and statistical processing errors, its validity, or at least of some experiments, has been questioned by some niches of the scientific community. A percentage of the software used to post-analyze the fMRI data have been found to contain faulty algorithms and are being fixed and updated. So, aware of these facts, I initiated a 4th series of experiments using a much more robust and simpler kind of NMR analysis called "proton spectroscopy." The results obtained so far are very interesting and support the findings described here and will be made public when possible.

As discussed, deepening the study of the possible mechanisms of interaction between bioenergy and the BOLD and other MRI techniques can inspire the development of better and more specific or dedicated systems for bioenergy detection as well as for the creation of an explicative and predictive theoretical framework for the conceptualization and understanding of bioenergy.

Nonetheless, the main consequences would not be in the areas of technology, convenience or human comfort, but in the areas of philosophy, holistic comprehension of a deeper nature of reality, and of our understanding of ourselves as multidimensional consciousnesses. Perhaps it would even reinforce other evidences that humans (and other living beings) are transcendent consciousness, so that, similar to the NDE (near-death experience), it can help people to expand their worldview, consciential maturity, personal principles, and collective ethics.

Endnotes

Aerts, D. & Sassoli de Bianchi, M. "Do spins have directions?" *Soft Computing*, DOI: 10.1007/s00500-015-1913-0 (2015).

Alegretti, Wagner (2008). "An Approach to the Research of the Vibrational State through the Study of Brain Activity"; Proceedings of the 2nd International Symposium on Conscientiological Research; *Journal of Consciousness* (formerly *Journal of Conscientiology*); Vol. 11, N. 42; International Academy of Consciousness (IAC); October 2008; Portugal; p. 221

Alegretti, Wagner (1990). "Tecnologia Bioenergética" (Bioenergy Technology); Proceedings of the 1st International Congress of Projectiology; IIPC; 1990; Rio de Janeiro, Brazil; p. 32

Pinheiro, Rute Maria Rodrigues (2013). *Correlato eletroencefalográfico do estado vibracional*; (dissertation for Physiological Psychology and Behavioral Studies master degree) - Universidade Federal do Rio Grande do Norte; Natal, Brazil; 2013

Trivellato, Nanci (2008). "Measurable Attributes of the Vibrational State Technique"; Proceedings of the 2nd International Symposium on Conscientiological Research; *Journal of Consciousness* (formerly *Journal of Conscientiology*); Vol. 11, N. 42; October 2008; Portugal; p. 168

Trivellato, Nanci (2017). "Vibrational State and Energy Resonance: Self-tuning to a higher level of consciousness"; International Academy of Consciousness; 2017; USA

Trivellato Nanci & Alegretti, Wagner (2005); "Bases para o Energograma e Despertograma"; conference presented during the I Jornada da Despertologia; CEAEC; 15 - 17 July 2005; Foz do Iguaçu, Brazil

Vieira, Waldo (2002). *Projectiology: A Panorama of Experiences of the Consciousness outside the Human Body*; IIPC; 2002; Rio de Janeiro, Brazil; p. 22

Some Reflections on Consciousness, Intention, and Healing

WILLIAM BENGSTON

In their thought-provoking piece "Is Consciousness the Life Force?,"[1] Dunne and Jahn propose, among many other ideas, the possibility that

- Consciousness is inherently linked to the biological process of life itself
- If consciousness-related anomalies indeed prove to arise from an unconscious biological source, one would assume that such a correlation would serve a purpose advantageous to the development or survival of the organism
- The phenomena we term "anomalies" are mysterious primarily because they cannot be explained by any known physical model, but this well may be because they are not strictly physical...rather, they represent a domain where the objective physical world and the subjective experiential world overlap, and they hint at a connection to the primal Source from which all phenomena originate
- Their thesis is that the body, through the agency of the life force, may represent that intersection through a more direct connection with the Source than the mind, or at least the conscious mind...

For over 35 years, I have been researching anomalous healing, using both *in vitro* and *in vivo* models. My research agenda has looked at some parameters of healing such as distance and dose; the physiological correlates of healing; and more recently, the attempt to reverse engineer the healing effect to make it a scalable and reliable more conventional treatment.

Sixteen *in vivo* cancer experiments on mice have been conducted using standard models of mammary adenocarcinoma, methylcholanthrene-induced sarcomas, naturally occurring oncogenic tumors, immune-deficient nude mice, and on an extremely aggressive cancer. These have been performed at eight different institutions, including four medical schools. In addition, innumerable in vitro experiments have been performed on human leukemia cells and human breast cancer cells, to name but a few.

The experimental protocol used in all of these experiments has been to take a "standard" mouse or cell model with a long history of conventional empirical research that has a known and predictable outcome,[2,3,4] and to introduce the variable of "healing with intent" using a healing technique which I helped to develop.[5] Volunteer healers, who have included both students and faculty, have all been pre-screened to be without any experience in alternative healing, nor were they in any way "believers" in the validity of alternative healing. Their healing experiences with me were the first of any kind for them. Variations in many parameters have been examined, such as distance and dose and frequency of treatment, as well as the compilation of the subjective experiences of the volunteer healers. Additionally, human physiological correlates have been examined using EEGs at a private lab, and fMRIs independently carried out at two medical schools.

The abridged summary of the results of these experiments include:

- The demonstration of a reliable full lifespan cure of cancer in experimental mice, including an apparent immunity to reinjection of the same cancer
- A dose response to healing, with some minimum amount of healing time being necessary to affect a cure. Interestingly, the only predictor of the aggregate speed of cure is the number of mice in an experiment, the quicker cures being associated with *more* mice being treated
- Healing proceeds in non-linear fashion, with sudden bursts of healing that resemble "phase transitions"
- A measurable "resonant bond" between healer and healee that is fluid, with successful healing being associated with "connection" and healing failure being associated with "disconnection"
- Healing appears to be fundamentally about "information" rather than "energy," despite the popular use of the latter term.

Perhaps most appropriate for the following discussion on the place of consciousness in all this, healing appears to be unrelated to a particular conscious state of awareness on the part of either the healer or the healee, and healing can apparently be stored in both biological and physical systems. That is,

- Healing efficacy is not necessarily related to a particular state of mind, but rather appears to be more akin to an autonomic response to need
- There is a measurable resonant bond between the brains of the healer and healee that can be strengthened or weakened
- Cells transplanted from a mouse infected with cancer that has been treated by the healing technique can independently cure otherwise infected mice, without further healing with intent
- Water treated with healing with intent can reproduce the cancer cures in mice without further intervention on the part of a healer
- *In vitro* cancer cells treated with cell medium that has been "charged" with the healing with intent will result in a robust acceleration of growth
- *In vitro* cancer cells treated with "charged" cotton will undergo significant genomic changes related to immunology and inflammation
- A digital "recording" made of the act of charging cotton, when played to cancer cells, will produce significant genomic changes similar to that of charged cotton
- And finally, these apparent effects of healing, whether through healing with intent, charged cotton, or a recording, produce significant responses *only when there is a biological healing need present in the healee.*

My discussion will move in the direction of suggesting that (1) healing is not related to conscious awareness, but rather is more akin to an autonomic biological response to need; (2) conscious intention and attention are quite dissimilar, with only fleeting intention being the necessary component of healing; and (3) intention itself may be "storable" in some form, and can result in relatively permanent changes in both biological and physical systems.

The end result of the discussion will essentially be in support of Dunne and Jahn's perspective that biology, more specifically biological "need," may be the driving force behind a great deal of what can be considered anomalous healing. The healee will be seen to be the propagator of healing, drawing upon the intention of the healer to stimulate the process. As with so many other biological processes, the driving force is need rather than conscious awareness.

How consciousness fits into all of this will be the "wild card." While anything approaching a comprehensive definition of consciousness remains past my pay grade, I will suggest that a more useful conceptual model flows from the distinction between "awareness" and intention," the latter being the dominant partner and the former being optional at best.

By extension, the thrust of my data will be in support of the M5 conceptual model of Jahn and Dunne,[6] in that the mysterious "Source" will be the place which connects conscious intention and anomalistic outcome in healing.

An Illustration of the *In Vivo* Experimental Model

In the basic experimental *in vivo* protocol, mice are injected with a known dose of cancer sufficient to guarantee death within a specified interval. In the illustration below, mice obtained from either the Jackson Laboratories or the National Cancer Institute are injected with at least 200,000 mammary adenocarcinoma cells, double the lethal dosage. Published life expectancy has been found to be between 14 and 27 days subsequent to injection. Mice develop a non-metastatic externally palpable tumor that results in death either by the crushing of the internal organs or by malnutrition, or both. An example of a mouse very close to death can be seen in Figure 1.

Figure 1 – An example of a mouse in the end stages of life

Healing treatment of the mice generally involves the volunteer healer placing his or her hands on the outside of the cages and practicing the healing technique for a specified duration (see Figure 2). Individual

Figure 2 – A typical healing session, with the author serving as the healer

treatment length, number of treatments, and number of mice per treatment have all been part of various experimental protocols, including other variables such as distance of hands from the cages, extending to thousands of miles.

Those mice that have been treated by the healing with intent techniques typically develop an encrusted blackened area on the surface of the tumor, followed by tumor ulceration and then implosion to full lifespan cure (see Figure 3).

Histology indicates that at all stages of remission there are viable cancer cells in the mouse. When full cure is achieved, the mouse is completely free of cancer, and is further apparently immune to subsequent injections of the same cancer for its entire life.

The sudden disappearance of the cancer can also be illustrated in Figure 3. The first and second pictures are spaced only six days apart. Sudden shifts, analogous to "phase transitions," have been the pattern in all experiments, whether *in vivo* or *in vitro*. That is, in early stages of healing treatment, mice (and cell cultures) show no apparent effect of the healing intention, until suddenly there are non-linear dramatic shifts in tumors (*in vivo* models) or cell growth (*in vitro* models).

Figure 4 illustrates the sudden growth in cells in response to being grown in medium that has been charged with the healing with intent method. After the first week, there is no significant difference in cell growth between treated and untreated

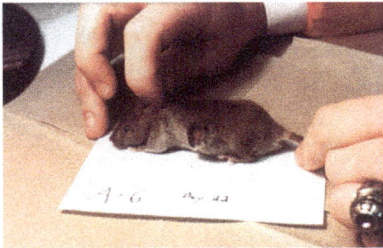

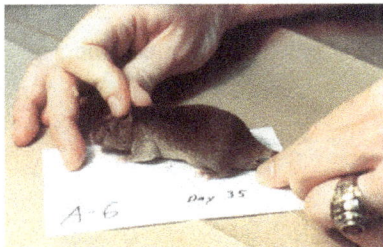

Figure 3 – Illustration of the remission pattern

medium. But in the second week there is a sudden burst in growth in cells grown in the treated medium.

After the first week, there is no significant difference between cells treated and untreated. After the second week, the cells grown in the treated medium are significantly stimulated.

Healing and the Sense of "Connection": EEG data

In addition to the anomalous healing in and of itself, there is apparently an anomalous connection that can occur between subjects.

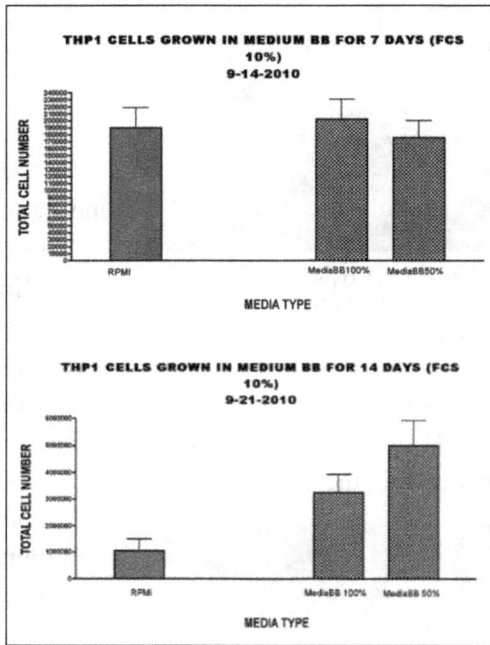

Figure 4 – a treated/not treated medium comparison of cell growth in human leukemia cells at one and two weeks

Synchronizing EEGs between the healer and subject indicate that the healer's EEG data shows harmonic frequency coupling across the spectra, followed by frequency entrainment effects with the healee, followed by instantaneous EEG phase locking. These results suggest the presence of a connection between the healer and healee (see Figure 5).[7] Of central importance, in addition to the apparent connection established at a distance, is that *neither the healer nor healee are consciously aware of the connection.* The healee has a need, and the healer practices the rapid imaging healing technique with only a passing intention to help. The connection simply occurs.

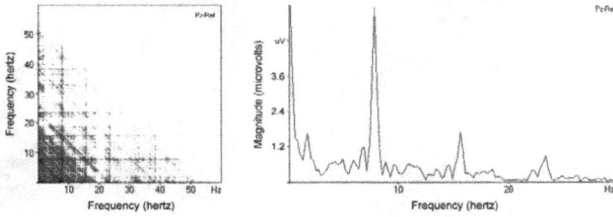

7.81 Hz spindle with harmonics at 15.63 and 23.44 Hz, bispectra and spectra.
   (A) is the bispectral display showing the cross-spectral interaction of the 7.81 Hz EEG activity and the other mathematically related harmonics, seen as a "checkering" of the display.
   (B) displays the spectral magnitude peaks, seen with the 7.81 Hz primary peak, and then the progressively smaller magnitude responses as the doubling and tripling harmonic peaks are seen. Peak values should be interpreted as midpoints of possible values of a given resolution. In this case, 7.81 is the midpoint of possible values within .125 Hz.

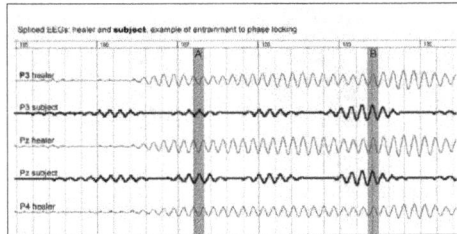

Spliced EEGs: healer and subject, example of entrainment to phase locking.
   The figure shows waveforms revealing the 7.5 to 8 Hz frequency range in both healer and subject at three parietal locations. Early in the healer's sustained amplitude burst (A), the phase of the subject does not match the healer's. As the healer's burst continues, the subject's phase synchronizes with the healer's (B) as the subject's amplitude also increases to near its maximum for the entire 11-min session.
   Note: Given the healer's greater amplitude generally, the subject's entire waveform was amplified for clarity (50 vs. 15 microvolts/cm sensitivity).

Figure 5b – Illustrations of the production of harmonic frequency coupling

Healing and the Sense of "Connection": fMRI data

Research questions have included whether healing with intent might have a specific location in the brain. In order to gather data using functional MRIs, it would be necessary to produce a "toggling" of healing intention into "on" and "off" states. To my amazement, this was able to be done

Figure 6 – fMRI data contrasting "on" and "off" healing intention, with a control run

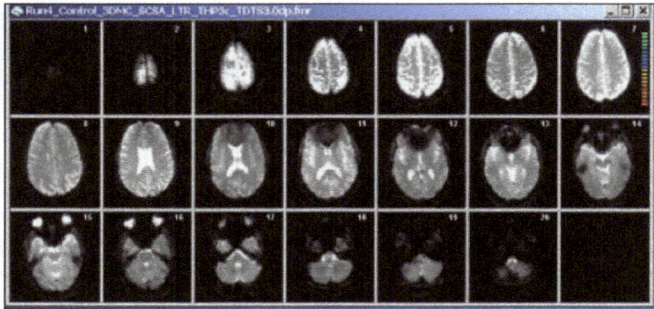

This is a control run with a key-pressing task without on/off cueing. It shows no activation when 45-second blocks were contrasted.

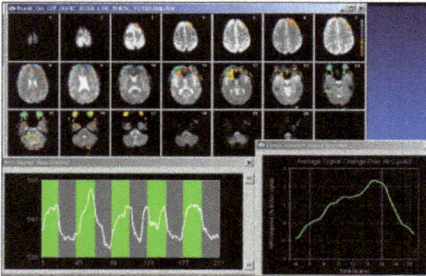

fMRI data show widespread activation when healing is "on" (green) as contrasted with "off" (grey). Right anterior inferior frontal lobe is highlighted as region of interest and shows ~3% signal increase.

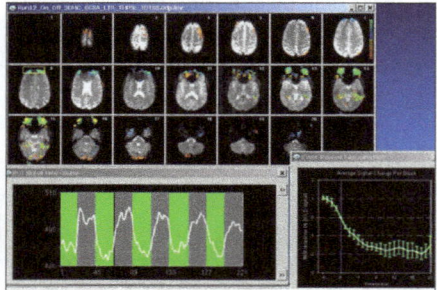

Anterior frontal lobes highlighted at an inferior level on the region of interest. 5 periods are averaged showing ~3% decrease during "on."

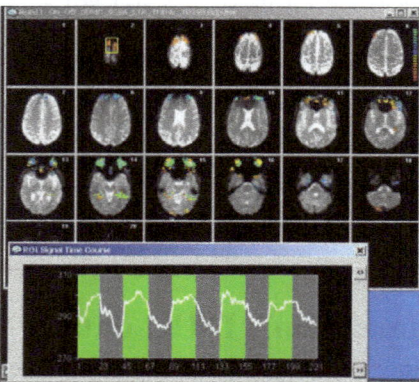

fMRI data show widespread activation when "on" (green) as contrasted with "off" (grey). Anterior frontal lobes are highlighted as the region of interest.

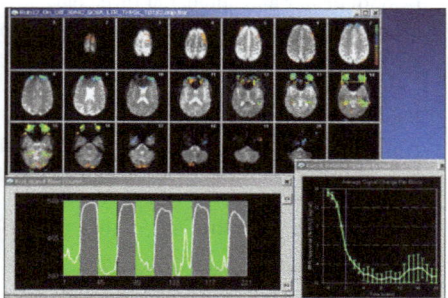

Both eyes highlighted as region of interest, and 5 periods are averaged showing ~25% signal decrease during "on" conditions.

through conscious intent. At the University of Connecticut and Thomas Jefferson Medical Schools, an exploratory pilot study with a simple protocol had healers inside an enclosed fMRI intend to "heal" and then to "not heal" during 45-second cycles (see Figure 6) to see if healing intention can be toggled.

Apparently, healing can indeed be "toggled." These results were reproduced by several people acting as healers, always contrasting "on" and "off" states of healing intention, using the same techniques as were applied to the mice and in the EEG studies.

An interesting modification of the protocol involved the healer standing outside the fMRI, approximately 25 feet from the fMRI, and a volunteer healee located inside the fMRI. The healee had no specific intention, and was instructed to simply lie inside of the fMRI. Here too, the healer was the one cued to direct healing intention in an on/off cycle of 45 seconds each, except this time the healee was the one being monitored. *The same basic pattern of on/off cueing in the brain of the healee was produced, apparently indicating a brain connection across some distance. Once again, there was no conscious awareness on the part of the healee that anything was out of the ordinary. The healee's only task was to lie still inside the fMRI.*

The third variant on the fMRI protocol involved the gathering of 10 pictures and hair samples of cancerous animals (there were dogs, cats, horses, and sheep) which were each placed inside of an opaque envelope. To serve as controls, an equal number of opaque envelopes were prepared that had only index cards inside of them. The envelopes were randomized, and in double blind fashion, were each placed on the left palm of volunteer healers who were lying in an enclosed fMRI. *Results clearly indicate that the brains of the volunteer healers "turned on" only when the envelopes had "need" expressed in them (pictures and hair samples of cancerous animals). This apparent activation in response to need essentially duplicated the results when the healers consciously attempted to toggle healing and non-healing in specified intervals. The brains "knew" when a need was present in the envelopes.*

Healing and Connection: The Control Problem

The previous sections indicated that bonding between spatially separate individuals is not necessarily a conscious process. In both the EEG and fMRI experiments, brains apparently either became resonant with each other or autonomically responded to the stimulation of healing need. This bonding further occurred without the conscious awareness (read "attention") of the participants, however much it was their "intention" to participate in the experimental protocols.

Among the more interesting phenomena associated with the healing research has been the persistent complication of control mice remissions. The current interpretation is that mice that are even briefly seen by the volunteer healers can somehow become "bonded" to the treated mice. The effect of this bonding is that a treatment to any mouse apparently can result in a treatment given to any mouse within the bonded mice system. Of equal importance, resonant bonds are apparently fluid, in that bonds can also be broken.

To illustrate, in Figure 7 (below), five cages of mice on a lab bench were treated by five volunteer students. Three of the volunteers were biology students, while two were not. The students were to treat their cages each day for a specified length of time. They were also told *not* to look for the room with the control mice in them. This control room was in the same building about 50 meters away. A second set of controls mice were shipped to another city.

Several weeks into the experiment, the control mice in the building started to die on schedule. When the biology students heard this, they defied instruction and went to find the control mice in the building, rationalizing that they would just briefly peek at them and then leave. They found the control mice to be in similar state as shown in Figure 1 (above), without any blackened area or ulceration of the tumor. They stayed only about 10 minutes to observe. After their visit, the remaining control mice began the remission process to full cure, even though they were never again seen by the volunteer healers.

The end result of this part of the experiment found the non-biologists' mice being cured, while the biologists' mice died. The control mice were dying on schedule until they were seen by the biologists, who apparently couldn't cure their own mice!

And so the problem is this: if "connection" is part of healing (see EEG and fMRI data above), and the biology students were able to resonantly connect to the control mice to apparently cure them, why couldn't they cure their own mice? Further, if the non-biologists were able to cure their assigned mice, and they were close to the biologists' mice, why wasn't there a resonant bond made with the biologist's mice?

A hint comes from their student logs. Each of them reported feeling self-conscious about whether they would be seen doing something so unorthodox as putting their hands around a cage of mice, thereby exposing themselves to ridicule by their peers. The non-biology students had no such fear. Apparently, the subjective state of unease can serve to break the resonant bond with the larger group. This particular state of consciousness, simply described as self-consciousness, might be one of a possible multitude of states which can unbond.

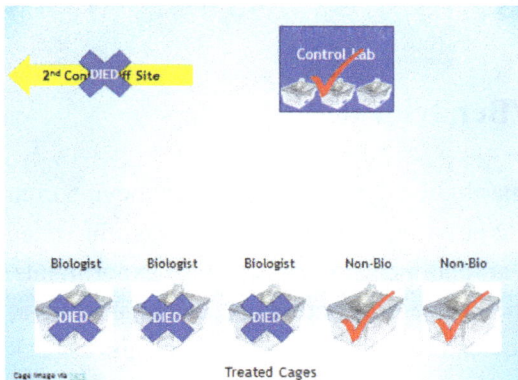

Figure 7 – experimental layout, with outcomes, of the lab setting

This can be further illustrated by an additional quirk to this experiment. The experimental protocol pushed some IACUC boundaries in that each of the five volunteer healers also took a cage of mice home, and each of them was the only person to see and treat their home mice.

All of the home mice were cured. Even the biologists were able to cure their mice at home, where they were, by their own logs, more relaxed. In their logs, they also report being excited by their discovery of the control mice in the building, which presumably served to resonantly bond with the remaining mice in the larger group.

These results don't make sense if there is anything approaching a generic "field effect" of healing. If the non-biologists could heal in the lab, and the biologists' mice were in the same vicinity, then a field effect would have compensated for the biologists' apparent lack of being able to heal in the lab. Apparently, states of consciousness which "push away" might serve to create boundaries which exclude healing.

The resonant bonding problem, then, isn't one of a generic field effect, but rather involves the interaction of consciousness to bound group membership. This would indicate that models such as morphogenetic fields,[8, 9] however applicable in general terms, might need consciousness as an intervening variable in specific instances.

Interestingly, if the resonant bonding phenomenon is more generally widespread and not simply confined to healing data, then a similar process may be involved in placebos.[3] That is, placebo effects may be less explained by psychological processes, and more explained by sociological processes of group bonding and boundary formation and dissolution.

Need and a Biological/Behavioral Response to Healing in Mice

Among the curious phenomena that have been observed is the apparent response of mice to the healing environment. In Figure 2 was illustrated the physical placing of hands around a cage of mice. In all experiments, we have seen a preference by the mice to situate their tumors as close as possible to the left palm of the healer (see Figure 8). This occurs regardless of the orientation of the cage.

This left-handed attraction occurs regardless of the healer, or the type of cancer, *only so long as the mice have cancer.* Once the mice have been completely cured, they no longer have an inclination to move towards the left hand. If the pictures in Figure 8 were transformed into motion, it would appear as if the mice actually rotate their turns in their attempt to

Figure 8 – apparent left-hand attraction in two different mouse models

get the tumors as close as possible to the left palm, and they switch off after approximately 1 minute of minimum distance from that placement.

Healing and Need in Cell Cultures

Ordinary cotton obtained from a pharmacy (Figure 9) was "charged" by a volunteer healer for approximately 20 to 30 minutes. Charged and uncharged (control) pieces of cotton were then placed besides well plates with cells that have "need" and cells without need.

Several growth experiments were done on bacteria cells without healing need, comparing the effects of "treated" and "untreated" cotton. There was no significant effect on cell growth (see Figure 10).

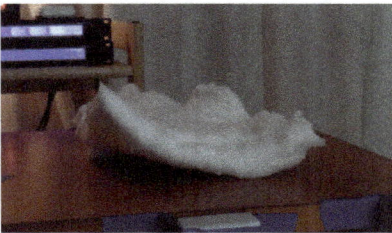

Figure 9 – Ordinary cotton treated with healing intent

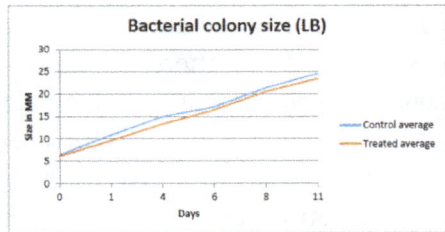

Figure 10 – Proteus vulgaris bacteria (luria broth agar) growth experiment of cells without healing need

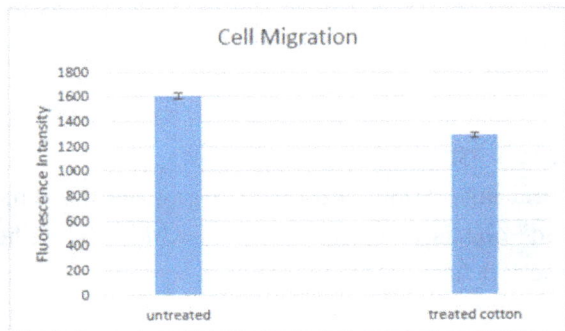

Cotton treated cells demonstrated slower cell proliferation and migration

Figure 11 – Human breast cancer cell proliferation and migration comparisons between charged and uncharged cotton

Human Breast Cancer cells exposed to Energized Cotton (4 replicates)			
Gene Description	Gene Symbol	Energized cotton	
		Fold Change	P-Value
Caspase 9, apoptosis-related cysteine peptidase	CASP9	-1.241	0.017
E2F transcription factor 4, p107/p130-binding	E2F4	1.123	0.023
Heme oxygenase (decycling) 1	HMOX1	-1.310	0.034
Insulin-like growth factor binding protein 3	IGFBP3	1.181	0.016
Minichromosome maintenance complex component 2	MCM2	1.320	0.020
Protein phosphatase 1, regulatory (inhibitor) subunit 15A	PPP1R15A	-1.435	0.008
Serpin peptidase inhibitor, clade F (alpha-2 antiplasmin, pigment epithelium derived factor), member 1	SERPINF1	1.406	0.083
Vascular endothelial growth factor C	VEGFC	1.218	0.093

Figure 12 – Significant genomic changes in human breast cancer exposed to charged cotton

However, when human breast cancer cells with a healing "need" were exposed to the treated vs. untreated cotton, significant changes occurred in cell proliferation and migration (Figure 12).

Genomics performed on human breast cancer cells, comparing exposure to treated and untreated cotton, demonstrated significant changes in six genes if the cotton was charged, with strong suggestion of an additional two genes which may be of importance.

Interestingly, Raman spectra analysis failed to detect any difference between the charged cotton that had produced genomic effects, and uncharged cotton. The testing of other materials besides cotton included clear quartz, pink quartz, water, selenite, and cellulose, all with no detectable difference between "charged" and "uncharged."

To date, only life in need seems to respond to charged materials. And so the interesting speculative hypothesis might be that *the detector for the difference between healing and non-healing might need to be both alive and in need.*

Some Concluding Thoughts

Several interesting patterns emerge from these data. First, the theme of "need" consistently occurs. Mice which have a healing need will move to the left hand of the healer. Once they are completely cured, this no longer happens. Similarly, cells which have a healing need will respond to healing with intent, whether that healing source comes directly from the hands of a healer, or healing apparently stored in substances such as water, cell medium, or cotton. Cells which have no healing need exhibit no anomalous changes when offered healing with intent. And so, at a minimum, it can be posited that biological need is a crucial component in healing, and it may be the healee which instigates the healing effect.

Second, conscious awareness on the part of the either the healer or healee is not likely to be central to produce a healing effect. The extent or quality of consciousness on the part of mice or cells may be debatable, but there is little question that they are unlike anything that parallels human consciousness. Yet mice in need "know" to move proximate to a healing source; cells in need do likewise. Can there be serious doubt that these responses to healing are natural biological responses?

The volunteer healer logs vary widely in the extent to which they were consciously aware of anything associated with healing. Some occasionally felt some sort of "connection" with their mice; some felt nothing at all, the latter to the point that they seemed not to understand the question when asked to comment about feelings associated with healing. In multiple experiments, there has been no association between healing efficacy and subjective states of connection. And so it may be posited that a conscious awareness of healing may be unnecessary for healing to take place.

At the same time, both EEG and fMRI data clearly indicate that some sort of biological connection actually does take place. For one, at least in the case of healing humans, healer and healee go into harmonic brain phase locking, and do so without any necessary conscious awareness that the healing phenomenon might be taking place.

Yet, while there can be high confidence that conscious *awareness* on the part of either the healer or healee is optional at best, the role of *intention* on the part of the healer becomes more problematic. That is, the simple act of putting hands around a cage, or attempting to "charge" materials for a healing experiment, signifies intention of some sort. That intention may be fleeting, and certainly separate from anything approaching either belief or sustained awareness. But if action is taken in order to produce or to test healing of any sort, there must be intention. This intention is akin to the intention expended for many forms of action. I "intend" to walk down the street, but there is nothing approaching "belief" or sustained "attention." Indeed, attentive walking will diminish efficacy. There must have been some intention to begin the walking, but the activity is driven not by sustained effort or attention, but by a letting go. The mastery of many skills, whether walking or healing, likely involves the transition from "mindful" attention to relatively "mindless" fleeting intention.

Connection can be seen as an autonomic response to need. Consider that in one fMRI protocol, blinded envelopes with pictures and hair samples of animals placed onto the palm produced significant brain response in the healer if "need" was present in those envelopes. These responses were biologically similar to the brain changes which occurred if the healer intentionally attempted to heal. If the envelopes placed into the palms of volunteer healers did not contain need, then no brain changes ensued. Again, there was no conscious awareness of whether any envelopes did or did not have any pictures of animals in need.

The commonly found association between certain states of consciousness and healing, often associated with being "spiritual," likely has the temporal sequence inverted. Instead of a "spiritual" sense of connection being necessary in order to produce healing, the data indicate that healing occurs more as an autonomic response to biological need, and the subjective sense of spiritual connection is an optional consequence of that need. Since more subjectively sensitive individuals are more likely to be drawn to healing, the mistaken association can be made that this sensitivity is the source of healing. It turns out that people who are less subjectively sensitive can heal just as well without ever experiencing connection. Conscious awareness of spiritual connection is optional.

That healing intention can apparently be stored in materials and later be used to stimulate healing effects is extremely suggestive that consciousness may have an associative technology. Data presented here on the apparent storage of healing in water, cell medium, and cotton, which can produce a future healing effect when need is present, begs for inquiry into future studies that may help to unravel some of the mysteries of healing. And, there is the additional possibility that this storage ability might be able to make healing more conventional and scalable.

Finally, the lack of necessity of awareness of spiritual connection on either the part of the healer or healee makes it likely that healing does not conform to models of psychokinesis that support conscious intention as the operative agent. That is, healing outcome is not "willed" in the way that operators can bring about intended alterations in, say, the theoretical output of random number generators. Volunteer healers may have "intended" healing in that they went through some training in a healing technique and placed their hands around cages, but there are no cases of these healers having healing follow their wishes. Indeed, initial experiments proceeded under the assumption that if healing were to work, then mice that were treated shortly after being injected with cancer would avoid tumor growth altogether. In all cases, regardless of type of cancer, and regardless of how soon after injection treatment began, tumors grew, sometimes very large, before the process of ulceration and implosion commenced. The volunteer healers were successful in the outcome; they were upset and concerned when their mice developed tumors. Certainly, the pattern and stages of healing does not conform to the wishes of the healers.

The data output from the experiments, and the experiences of the healers, do not conform to anything like a direct PK effect. Instead, there is merit to thinking of healing as a non-directed outcome similar to that proposed by Jahn and Dunne in their M5 model for explaining their consciousness-related anomalies with random event generators and remote perception studies.[6] In that model is proposed the notion that the conscious mind might connect to the tangible physical world not directly, but by way of a circuitous route involving unconscious processes and intangible physical mechanisms. Further speculation involves a timeless and spaceless "Source" in which the unconscious and intangible merge.

While a full examination of the application of the M5 model to healing is beyond the scope of this chapter, let me simply say that the actual healing technique used in these experiments,[5] as well as the subjective experiences of a selection of volunteer healers is remarkably consistent with this model. This includes the speculative discussion of Source, which is directly discussed elsewhere.[11] I would add, however, that a full explication of the usefulness of the model to understanding healing would actually and controversially minimize the importance of the conscious mind.

Endnotes

1 Dunne, B. and Jahn, R. "Is Consciousness the Life Force?" This volume.

2 Bengston, W. and Krinsley, D., "The Effect of the Laying-On of Hands on Transplanted Breast Cancer in Mice." *J. of Scientific Exploration*, 14(3), pp.353–364, 2000.

3 Bengston, W. and Moga, M. "Resonance, Placebo Effects, and Type II Errors: Some Implications from Healing Research for Experimental Methods." *J. of Alternative and Complementary Medicine*, Volume 13(3), pp. 317–327, 2007.

4 Bengston, William F. "Spirituality, Connection, and Healing with Intent: Reflections on Cancer Experiments on Laboratory Mice." In Lisa Miller (ed). *Science and Spirituality*. Oxford University Press, 2012.

5 Bengston, William. "A Method Used to Train Skeptical Volunteers to Heal in an Experimental Setting." *J. of Alternative and Complementary Medicine*, 13(3), pp. 328-331, 2007.

6 Jahn, R. and Dunne, B. "A Modular Model of Mind/Matter Manifestations (M5)." *J. of Scientific Exploration*, 15(3), pp. 299-329, 2001.

7 Hendricks, L., Bengston, W., and Gunkleman, J. "The Healing Connection: EEG Harmonics, Entrainment, and Schumann's Resonances." *J. of Scientific Exploration*, Vol. 24, No. 4, pp. 655–666, 2010.

8 Sheldrake, Rupert. *A New Science of Life: The Hypothesis of Formative Causation*. J.P. Tarcher, 1981.

9 Sheldrake, Rupert. *The Presence of the Past: Morphic Resonance and the Habits of Nature*. Park Street Press, 1995.

10 Tiller, W. *Science and Human Transformation: Subtle Energies, Intentionality and Consciousness*. Pavior, 1997.

11 Bengston, W. "Hands-On Healing: A Training Course in the Energy Cure" (CD audio set). Sounds True, 2010.

Light, Biology, and Consciousness

BRENNAN KERSGAARD

In Norse mythology, the pursuit of knowledge and understanding is often expressed through tales of the god Odin. In one such story Odin journeys to the dimension-spanning world-tree known as Yggdrasil in order to imbibe from the sacred well at its base and receive eternal knowledge. Upon reaching his destination at the well, Odin is approached by its guardian, Mimir, who will only allow Odin to proceed and receive eternal knowledge under one condition. To drink of the well Odin must first gouge out one of his eyes. Odin ultimately acquiesces to this strange request and upon so doing is granted access to the waters of ultimate knowing, drinks and becomes enlightened. The crucial symbolism encoded within this story that is so appropriate to this chapter is that before Odin could reach ultimate knowledge, complete understanding of the whole cosmos and everything in it, he had to prepare his physical body to receive. He had to forsake the mundane perception of vision for a higher perception. He needed to learn how to see through a singular eye.

Seeing Through a Single Eye

When it comes to physiology we are used to speaking in pairs: two lungs, two kidneys, two eyes, etc. This complementarity is also present in many of the unpaired organs such as the left and right chambers of the heart, the left and right lobes of the liver, and the left and right hemispheres of the brain. Very few structures in the human body have a truly unitary nature. In the human brain, there is only one structure that is not polarized in any way. It is a very small mass of tissue, steeped in both ancient and contemporary mystical lore, buried deep in the center of the brain, known as the pineal gland. A cursory Internet search of the pineal gland will lead to an explosion of articles, blog posts and pictures about the pineal gland and "the third eye." But why is that so? How did this notion begin? Let us now, with sobriety and due diligence, unravel the hype and the folklore surrounding this fascination and pursue the relevant scientific and historical landmarks to better understand this enigmatic structure.

In 1865 a German anatomist by the name of Ludwig Stieda identified a small circular structure just below a patch of un-pigmented skin on the top of the skull of a small frog, on the exact midline of the skull just behind the eyes. Believing it to be a glandular structure he named it the frontal gland.[1] Less than a decade later another German anatomist identified the same structure on a lizard.[2] Unsure what to make of the find, contemporary zoologists dismissed it as a vestigial structure and went on with their studies.[3] Research eventually resumed in the 1930s when a team of researchers took to the powerful new electron microscope for a re-examination of this unusual structure. To their amazement, their new view of the structure revealed the presence of a fully developed lens and photoreceptors.[4,5] Subsequent electrophysiological experimentation would reveal that the structure was not vestigial at all, but capable of generating neural impulses when stimulated by light, much like the retina of the eye.[6,7,8] Once it was determined that this structure was not vestigial, the quest was on to discover its function. The earliest functional studies performed in lizards quickly revealed that this organ was helping its host synchronize its activity/rest circadian rhythm with the day/night cycle.[9,10]

It was at this point, approximately 60 years ago, that the reality of an anatomical "third eye" was first scientifically verified. It was soon discovered that this third eye was actually an emanation of the lizard pineal gland.[11] Instead of the frontal gland, as it was originally called, this new eye became known as the parietal eye, or sometimes the pineal eye. Many organisms, including lizards, frogs, fish and sharks contain both a photosensitive parietal eye and a pineal gland, joined together by the parietal nerve.[12] Together they are sometimes referred to simply as the pineal complex.

So that is the situation in many non-mammalian vertebrates, but what is the situation in those animals more closely related to man? Genetic and molecular studies have contributed significantly to our understanding of the mammalian pineal gland and its evolutionary origins.[13,14,15,16] For reasons that are still not altogether clear, over the course of evolutionary history the parietal eye was lost and the pineal gland began to recede deeper and deeper into the brain and into the dark, no longer sensing light directly, but instead relying on connections from the retina of the eye for its photonic input. That said, careful study has slowly revealed that even in mammals, the third eye lies latent within the pineal gland.

Some studies have shown that cells from the mammalian pineal gland (pinealocytes) transiently express photoreceptor features during certain phases of their development, features that are lost upon cellular maturation.[17] The extent of this genetic expression is so striking that some researchers have hypothesized that while the mature rodent pineal gland is definitely not photosensitive, the neonatal pineal gland likely is.[18] Well over a dozen studies, only a few of which are cited here, have shown that many of the key genes expressed in the mammalian retina necessary for vision are also highly abundant in the fully matured mammalian pinealo-cyte.[19,20,21] Mice lacking the ability to express a single gene crucial for ret-inal development known as OTX2 are unable to develop pineal glands.[22] These latest studies demonstrate a very strong functional, developmen-tal, and genetic bond between the two eyes all mammals use to navigate their visual environment and a third eye buried deep in their brain. In the words of one researcher:

> "Variations on a theme" seems to be a valid way to describe the pinealocyte and retinal photoreceptor. Although it is clear they have become specialized with respect to their relative capacity to…detect light, the concept that both evolved from a single photosensitive melatonin synthesizing cell seems obvious.[23]

And so it is that the third eye, an idea first borne out thousands of years ago, has been confirmed by modern science. This is remarkable and, perhaps for some, unexpected. But what is its ultimate meaning? Is it possible to see through your third eye? What would this mean? For this question, we will for a time depart the annals of academia and return to a more ancient wisdom.

A Spiritual Perspective

Many cultures have spoken of a third eye. In Hinduism, the third eye is associated with the ajna chakra and is believed to exist behind the center of the forehead.[24] The Hindu god Shiva is often depicted with a third eye in the center of the forehead. Hindus often paint a dot of paint known as a bindi between the eyebrows to represent the ajna chakra. The Buddha is likewise depicted in statues and in art with a third eye located between the eyebrows. In Taoism, disciples are trained to focus on the area between the eyebrows during meditative practice in order to exercise the third eye. In each case, learning how to "see" with the third eye is associated with spiritual illumination and profound insight.

In traditional Hindu and Buddhist spiritual practice the flow of subtle energies through the body is thought to coalesce into seven primary vortices, or chakras, running the length of the spinal cord. Each chakra is associated with certain developmental milestones and states of consciousness, and corresponds to different physical locations along the body. The 6th chakra is known as *Ajna* and corresponds to the center of the brain, behind the forehead. According to one 20th century yogic scholar and author, Swami Satyananda, as quoted in the article *Recent Research into A Possible Psychophysiology of the Yogic Chakra System*:

> The name Ajna comes from the root "to know" and "to obey and to follow." Literally the word Ajna means "command".... Yogis, who are scientists of the subtle mind, have spoken of telepathy as a "siddhi," a psychic power for thought communication and clairaudience etc. The medium of such siddhis is Ajna chakra, and its physical terminus is the pineal gland.[25]

While Swami Satyananda's views on the pineal gland may or may not be shared by his colleagues, it nonetheless demonstrates an interest in the pineal gland/third eye relationship by a well-known and published Hindu Swami. The English translation of Ajna as "command" is also worth discussion. The most extensively studied pineal hormone, melatonin, exerts a tremendously wide range of effects on human physiology. Melatonin receptors are prolific within the human body. They are found throughout the central nervous system as well as the gastrointestinal tract, lymphatic system, immune system, liver, lung, skin, adrenal gland, ovaries, testes, mammary tissue, prostate, skin, white blood cells, blood platelets, blood vessels and arteries, fat cells, kidney, heart, gallbladder, salivary glands, pancreas, and placenta.[26] Apart from acting in receptors, melatonin is also an incredibly powerful antioxidant,[27] capable of binding to important intercellular messaging proteins[28,29] and may assist in DNA repair.[30,31] With so many cellular and molecular actions, it is not surprising that melatonin and the pineal gland exert wide control over crucial organismal processes such as reproduction and puberty, protection from disease, digestion, energy balance, sleep, and cardiovascular health.[32,33,34,35,36,37,38,39]

Victorian theosophical practices also drew direct connections between the pineal gland and the third eye. In *The Secret Doctrine*, first published in 1888, Helena Blavatsky had this much to say, "The 'deva-eye' exists no more for the majority of mankind. The *third eye is dead*, and acts no longer; but it has left behind a witness to its existence. This witness is now the pineal gland."[40] This is a remarkable statement for two reasons:

1. The first scientific observation that the parietal or (pineal) gland of the lizard contained eye-like features such as a physical lens and photoreceptors occurred approximately 40 years after Blavatsky's original statement.
2. Her statement that "the third eye is dead but has left behind a witness" might also hint at the dynamic and mysterious evolution of the pineal gland from the surface of the skull to deep down into the center of the brain.

There are some reasons to believe that knowledge of the pineal gland and/or third eye was understood and encoded into traditional Christian cosmology. One tantalizing example of this comes from the Bible (Matthew 6:22) which reads, "The light of the body is the eye: if therefore thine eye be single thy whole body shall be filled with light."[41] Some have also suggested that the halos and rings of light often depicted in classical Christian art emanating from the heads of holy figures encode knowledge of the third eye and is another example of the center of the head being filled with light.[42]

Molecular Illuminations

Having thus established the validity of the third eye from both the scientific and the mystical perspective, we return again to a modern orientation and begin with a brief primer on evolutionary neurochemistry. And it starts with a single molecule, tryptophan. Tryptophan is an amino acid, which can be thought of as a letter in the protein-building alphabet. As the letters on this page are strung together to make words, amino acids are strung together to make proteins. Unlike the Latin alphabet, which uses 26 letters and makes words between 2 and 8 letters long on average, the protein alphabet uses 20 amino acids to make proteins hundreds of letters long. Tryptophan is the largest amino acid and the least often used, typically making up only 1–2% of the total weight of a protein. To really understand tryptophan, and its unique contributions to biology, however, one must zoom into its ringed core.

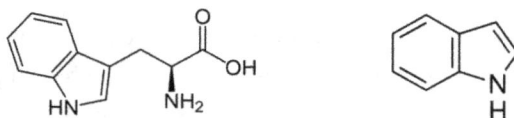

This structure is called indole and it has some remarkable properties. It is very rich in high-energy electrons, which when exposed to the correct frequency of light, are known to free themselves from their bonds and disperse into the surrounding environment with high probability.[43,44,45] This ability, known as fluorescence, helps tryptophan transmute solar energy into chemical energy useful for biological application. The profundity of this ability is difficult to understate, as the stripping away of electrons from donor molecules is the primary way that life fuels its myriad activities. In the words of one researcher:

> The creation of the indole structure served an important function in the start of aerobic life on the earth. The conversion of energy (photons) derived from the sun into biological energy requires capturing a light wave and the loss of an electron. Interestingly, the indole ring is the most efficient molecule for doing exactly this… Tryptophan was always a key to life because of its ability to convert solar energy into biological energy. The consequence

of this process made tryptophan and its associated molecules involved in all aspects of the organism's life.[46]

When the photosensitivity of indole combined with the protein-forming ability of an amino acid to form tryptophan, something truly larger than the sum of its parts was born. The relationship between tryptophan and light is as ancient as life itself and maintained into the present, and manifests itself in all the places you would think to look, such as crucial junctions in both photosynthesis and vision.[47,48,49,50,51,52] The situation regarding tryptophan and photons is tersely summed up by one author in the following words:

> "The capture of light by tryptophan is used at the active site by nearly all proteins (e.g., chlorophyll, rhodopsin and skin pigment cells) which capture light."[45]

Indeed, the relationship between tryptophan and light is highly intimate and very important to all life on this planet. But how does this relate more explicitly to the third eye and the pineal gland?

The bridge is found when one begins to realize the complementarity inherent within the structure of tryptophan. On the one hand, its unique ringed core makes it an essential, if not *the* essential link forever binding the domains of life and light together. And on the other hand, understanding that molecules built from tryptophan and retaining its ringed core are the neurological mediators of our perception of the cycles of light and darkness. Even more so, the neurological hub within which these tryptophan-derived molecules mediate our experiences of light and darkness is the pineal gland. These tryptophan-derived molecules are the previously discussed melatonin and to a lesser extent serotonin. So here we have two connections laying adjacent to each other. We have the physical structure of the melatonin molecule and the affinity of the indole ring for photonic energy, and also a functional role of melatonin to synchronize the body to the light/dark rhythm of the planet.

In summary, we now have enough clues to begin sketching a tentative three billion-year evolutionary arc detailing the interface of light and life. Beginning in the shallow primordial seas of a young planet there came about one of the earliest breakthroughs in the story of life—photosynthesis—propelled by the electrical properties and protein-building capacities of tryptophan. Photosynthesis is a demanding process and creates a very large free radical burden on the host cell. In these earliest photosynthesizers the need for antioxidative support would certainly have been at a premium. It is well documented that melatonin is a very ancient molecule and likely one of the earliest antioxidants utilized by the first unicellular photosynthesizing organisms to call Earth home.[53,54,55] As life evolved, so too would tryptophan continue to find ways to bring living creatures closer to the light, and serotonin and melatonin would go from humble antioxidants to important cellular signaling molecules, and eventually to possess widely celebrated and multifaceted roles in virtually all of our organ systems, including the gastrointestinal, cardiological, endocrinological, and nervous systems. Most importantly, these molecules, which have been so deeply connected with the sun's rays for over three billion years, occupy a central and commanding position within our own inner eye, the pineal gland.

The Psychedelic Pineal

Recent research is beginning to reveal the possibility that there is a third tryptophan-derived molecule beyond serotonin and melatonin, but very closely related in chemical structure, which may have its own distinct role to play in shaping our minds and bodies. This molecule is N,N-dimethyltryptamine, or DMT for short.

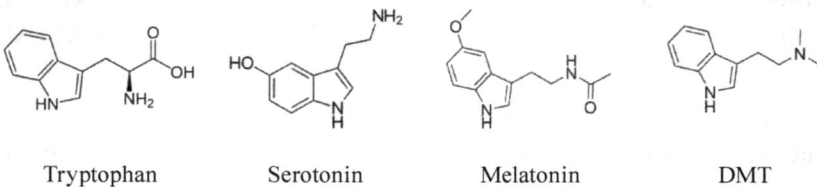

| Tryptophan | Serotonin | Melatonin | DMT |

Figure 1 – Tryptophan and its derivatives.

What is so startling about this is that DMT is one of the most powerful psychedelic drugs known in the world. In the 1990s, human laboratory research demonstrated that DMT was able to induce not only extremely rapid and intense perceptual shifts but, even more strangely, the sense of being visited from seemingly autonomous and intelligent entities in strange and surreal landscapes.[56,57] The profound mind-altering properties of DMT were first discovered scientifically in 1956, however ritual use of DMT containing plants is thought to go back thousands of years.[58,59] In 1965 a startling discovery was made when DMT was found in the blood and urine of healthy adult humans.[60] Since then over 50 studies have detected DMT in human bodily fluids and is now considered a normal constituent of human blood and urine.[61] More recent studies have also detected DMT in both whole rodent brain homogenate as well as the pineal gland.[62,63]

The above-mentioned studies prove that DMT exists within the human/mammalian body; however, they do not prove that DMT plays an active role in our physiology. For example, given the data thus far shared, it would be possible to argue that DMT is nothing more than a metabolic waste product of either melatonin or serotonin, and indeed this argument has been made. Research emerging within the last several years, however, is making this hypothesis easier to contest. For example, recent research has shown that DMT is not only actively transported across the blood/brain barrier but also actively transported into neurons from the intercellular milieu and stored inside the neuron for future use, leading some researchers to conclude that the absorption of DMT from the blood into the brain may indeed be a biological imperative.[64,65,66,67] DMT has also recently been shown to adaptively modulate the immune system when tested in cultured human white blood cells.[68] The obvious question is nicely summed up by the following researcher:

> There is no explanation as to why humans (as well as other animals) have evolved an endogenous compound to produce hallucinations, especially since there are no reasons to expect such false perceptions of reality to be adaptive.[64]

The surreal and profoundly visionary aspects of the DMT experience have garnered it great curiosity and enthusiasm among a wide range of media personalities and outlets, especially online. In fact, as this chapter

is being published, DMT has drawn itself right to the center of the renewed and thriving modern enthusiasm for the concept of the third eye
and its purported association with the pineal gland. One of the most interesting portrayals of DMT on the Internet that I am familiar with is a
website hosting over three hundred DMT "trip reports" or brief reports of
DMT users' experiences with the drug, what they saw, what their experience meant to them, etc. Here is one in its entirety:

> #17
> Suddenly, my world was filled with incredible multi-colored
> geometric designs that changed rapidly in a kaleidoscopic fashion.
> The visions were beautiful and ever changing. I can't ever recall
> seeing such colorful beauty with a psychedelic before.
>
> I opened my eyes, and the next thing I knew there was this face
> popping out of a book laying on my desk. It was difficult to make
> out, but it was clearly a female and she was smiling at me. Most
> dramatic were they eyes. They seemed to possess a life of their
> own. Unlike other psychedelics, this vision did not seem attached to
> my ego or me. Usually, I can see myself instilled in the surrounding
> objects, and everything appears to be an extension of my ego. I feel
> a connection to everything. Not this time: this face seemed to exist
> completely independent of myself. Was this an entity? Was this a
> connection? Whatever occurred, it was amazing to me. I shut my
> eyes again for a moment and when I opened them she was gone.

And passages from 2 others:

> #40
> I closed my eyes and was immediately face to face with a supreme
> being of sorts. It had no face, and its form was like a string of
> multicoloured lights constantly morphing; like a clown making a
> series of balloon animals, beginning with a dog, he alters a couple
> of parts, holds it differently to reveal a dolphin, and so on.

#44

There appeared in the vastness a tiny point of light. I remember realizing that I had not died at all, but that I had been dead. Then, not dead, but dormant. DORMANT. I was about to be born.

The feeling of flying is not an accurate description of the sensation that accompanied my movement toward the point, which was gold, and, to my surprise, was actually metallic. I came immediately upon the source, which was a DNA scarab, a construct, an insect of impossible dimensions, miles in diameter and circumference.[69]

Not all the user reports express this level of profundity; some do not detail any experiences with intelligent entities. However, there are remarkably consistent themes across the collection. A search of the entire collection for the word "light" found that a full 25% of this collection of 340 reports contained the word "light" within them in such a way as to suggest seeing or being surrounded by light during their DMT experiences.

Biophoton Emission

The apparent relationship between DMT and light as described in the above experiences, raises further questions regarding the role of light in consciousness and biology. Extensive studies of biophoton emission by German scientist Fritz-Albert Popp and others, indicate that light may play a central role in inter- and intra-cell regulation and communication.

Biophotons are photons that are emitted spontaneously by all living systems. It involves low luminescence, from a few up to some hundred photons per second per square centimeter surface area, at least within the spectral region from 200 to 800nm. They originate from a coherent (or/and squeezed) photon field within the living organism.[70] Popp suggests that biophotons may well provide the necessary activation energy for triggering all biochemical reactions in a cell. This light emission is an expression of the functional state of the living organism. For example, cancer cells and healthy cells of the same type can be discriminated by typical differences in biophoton emission.

It is hypothesized that the biophoton light is stored in the DNA molecules of the nuclei of the cells of the organism. The structuring and regulating activity of the coherent biophoton field produces a dynamic web of light that is constantly released and absorbed by the DNA to connect cell organelles, cells, tissues, and organs within the body. It serves as the organism's main communication network and its principal means for regulating all life processes, including morphogenesis, growth, differentiation, and regeneration. It has even been suggested that it may provide the basis of memory and other phenomena of consciousness.

How biophoton emission relates to the activity of DMT and the pineal gland is far from clear, but it is evident that light plays an important role in both processes. But as Popp has observed, "We know today that man, essentially, is a being of light."

Concluding Thoughts

The physical pineal gland is itself an enigma. It was the last gland of the human body to have its function identified. Yet it is the first gland to appear in the developing embryo, visible less than one month after conception. It is the smallest gland in the body and the only unpaired structure in the brain. For such a small structure, the pineal gland receives a tremendously rich blood supply. Gram for gram the only organ in the human body that receives a greater supply of fresh blood than the pineal gland is the kidney.[71]

Stories of a third eye span many cultures, and often such stories are associated with great wisdom, heightened intuition, as well as spiritual powers. In contemporary discussions on the matter, information abounds online in videos, articles, blog posts, cartoons, and artwork inspired by the third eye and its ancient lore. In modern times, the pineal gland, psychedelics and DMT are major components of the third eye excitement. One aspect that is seldom discussed in much detail, however, is the issue of light. I began my studies of the pineal gland, tryptophan, DMT, serotonin and melatonin because I wanted to understand the brain. In particular, to understand why our brains respond the way they do to psychedelics. So I travelled billions of years into the past and learned how melatonin contributed to both the regulation of the light/dark rhythms of single celled organisms as well as protected them from solar radiation induced free radicals as a powerful antioxidant.[72]

I came to recognize tryptophan's key position as a channel for photonic energy into cellular metabolism, enabling the biosphere to tap into the ultimate planetary energy source, the sun. And from there to read that 25% of DMT users experience seeing beautiful and bright light under the influence of the substance, to reflect on the old cultural adage of "seeing the light at the end of the tunnel" in reference to out-of-body travel and voyages to other realms, and to our own descriptive language when we speak of the bright light of heaven and the halos of saints. Our language is full of terms and phrases that show a deep reverence and respect for light. To say that an idea or concept is "illuminating" is the most direct of compliments. Another example is the word "insight," referring to an inner, mental vision of things. We speak so naturally of an inner vision of things, illuminating mental landscapes. Whether this inner vision actually exists or is only a useful metaphor, it is nonetheless fascinating that lying in the center of our brains is an inner eye. Could this inner eye be a physical terminus for an altogether different type of vision? It may yet be so.

Endnotes

1 Stieda, L. (1865)."Über den Bau der Haut des Frosches (Rana temporaria)." *Arch. Anat. u. Physiol.*

2 Leydig, F. (1873). "Ueber die äusseren Bedeckungen der Reptilien und Amphibien." *Archiv für mikroskopische Anatomie*, 9(1), 753–794.

3 Eakin, R.M. (1970). "A Third Eye: A century-old zoological enigma yields its secrets to electron-microscopist and neurophysiologist." *American Scientist*, 73–79.

4 Eakin, R.M. & Westfall, J. A. (1959). "Fine structure of the retina in the reptilian third eye." *The Journal of biophysical and biochemical cytology*, 6(1), 133–134.

5 Eakin, R.M. & Westfall, J. A. (1960). "Further observations on the fine structure of the parietal eye of lizards." *The Journal of biophysical and biochemical cytology*, 8(2), 483–499.

6 Miller, W.H. & Wolbarsht, M.L. (1962). "Neural activity in the parietal eye of a lizard." *Science*, 135(3500), 316–317.

7 Hamasaki, D.I. (1968)." Properties of the parietal eye of the green iguana." Vision research, 8(5), 591–599.

8 Dodt, E. & Scherer, E. (1968). "Photic responses from the parietal eye of the lizard Lacerta sicula campestris (De Betta)." *Vision Research*, 8(1), 61–72.

9 Stebbins, R C. & Eakin, R M. (1958). "The role of the ' third eye' in reptilian behavior." *American Museum novitates*; No.1870.

10 Glaser, R. (1958). "Increase in locomotor activity following shielding of the parietal eye in night lizards." *Science*, 128(3338), 1577–1578.

11 Engbretson, G.A. & Lent, C.M. (1976)." Parietal eye of the lizard: neuronal photoresponses and feedback from the pineal gland." *Proceedings of the National Academy of Sciences*, 73(2), 654–657.

12 Dodt, E. (1973). "The parietal eye (pineal and parietal organs) of lower vertebrates." In *Visual Centers in the Brain* (pp. 113–140). Springer Berlin Heidelberg.

13 Klein, D C. (2006)." Evolution of the vertebrate pineal gland: the AANAT hypothesis." *Chronobiology international*, 23(1–2), 5–20.

14 Falcón, J., Besseau, L., Fuentès, M., Sauzet, S., Magnanou, E., & Boeuf, G. (2009). "Structural and functional evolution of the pineal melatonin system in vertebrates." *Annals of the New York Academy of Sciences*, 1163(1), 101–111.

15 Mano, H. & Fukada, Y. (2007). "A Median Third Eye: Pineal Gland Retraces Evolution of Vertebrate Photoreceptive Organs." *Photochemistry and photobiology*, 83(1), 11–18.

16 Klein, D. C. (2004). "The 2004 Aschoff/Pittendrigh lecture: theory of the origin of the pineal gland—a tale of conflict and resolution." *Journal of biological rhythms*, 19(4), 264–279.

17 Zimmerman, B. L. & Tso, M. O. (1975).."Morphologic evidence of photoreceptor differentiation of pinealocytes in the neonatal rat." *The Journal of cell biology*, 66(1), 60–75.

18 Blackshaw, S. & Snyder, S.H. (1997). "Developmental expression pattern of phototransduction components in mammalian pineal implies a light-sensing function." *The Journal of neuroscience*, 17(21), 8074–8082.

19 Beere, P.A., Glagov, S. & Zarins, C. K. (1983). "Rhodopsin kinase activity in the mammalian pineal gland and other tissues." *Psychosom. Med*, 45, 95.

20 Korf, H.W., Møller, M., Gery, I., Zigler, J.S. & Klein, D.C. (1985). Immunocytochemical demonstration of retinal S-antigen in the pineal organ of four mammalian species." *Cell and tissue research*, 239(1), 81–85.

21 Korf, H.W., White, B.H., Schaad, N.C., & Klein, D C. (1992). "Recoverin in pineal organs and retinae of various vertebrate species including man." *Brain research*, 595(1), 57–66.

22 Nishida, A., Furukawa, A., Koike, C., Tano, Y., Aizawa, S., Matsuo, I. & Furukawa, T. (2003). "Otx2 homeobox gene controls retinal photoreceptor cell fate and pineal gland development." *Nature neuroscience*, 6(12), 1255–1263.

23 Klein, D.C. (2006). "Evolution of the vertebrate pineal gland: the AANAT hypothesis." *Chronobiology international*, 23(1–2), 5–20.

24 Saraswati, S.S., & Nikolić, N. (1984). *Kundalini tantra*. Bihar, India: Bihar School of Yoga.

25 Roney-Dougal, S.M., *Recent Research into A Possible Psychophysiology of the Yogic Chakra System*.

26 Ekmekcioglu, C. (2006). "Melatonin receptors in humans: Biological role and clinical relevance." *Biomedicine & Pharmacotherapy*, 60(3), 97–108.

27 Tan, D.X., Manchester, L.C., Terron, M.P., Flores, L.J. & Reiter, R. J. (2007). "One molecule, many derivatives: A never-ending interaction of melatonin with reactive oxygen and nitrogen species?" *Journal of pineal research*, 42(1), 28–42.

28 Benitez-King, G., Huerto-Delgadillo, L. & Anton-Tay, F. (1993). "Binding of 3 H-melatonin to calmodulin." *Life sciences*, 53(3), 201–207.

29 Romero, M.P., García-pergañeda, A., Guerrero, J.M., & Osuna, C. (1998). "Membrane-bound calmodulin in Xenopus laevis oocytes as a novel binding site for melatonin." *The FASEB journal*, 12(13), 1401–1408.

30 Sliwinski, T., Rozej, W., Morawiec-Bajda, A., Morawiec, Z., Reiter, R. & Blasiak, J. (2007). "Protective action of melatonin against oxidative DNA damage—chemical inactivation versus base-excision repair." *Mutation Research/Genetic Toxicology and Environmental Mutagenesis*, 634(1), 220–227.

31 Liu, R., Fu, A., Hoffman, A.E., Zheng, T. & Zhu, Y. (2013). "Melatonin enhances DNA repair capacity possibly by affecting genes involved in DNA damage responsive pathways." *BMC cell biology*, 14(1), 1.

32 Mori, N., Aoyama, H., Murase, T. & Mori, W. (1989). "Anti-hypercholesterolemic Effect of Melatonin in Rats." *Pathology International*, 39(10), 613–618.

33 Holmes, S.W., & Sugden, D. (1976). "Proceedings: The effect of melatonin on pinealectomy-induced hypertension in the rat." *British journal of pharmacology*, 56(3), 360P.

34 Csaba, G., Bodoky, M., Fischer, J. & Acs, T. (1966). "The effect of pinealectomy and thymectomy on the immune capacity of the rat." *Experientia*, 22(3), 168–169.

35 Markus, R.P., Cecon, E., & Pires-Lapa, M.A. (2013). "Immune-pineal axis: Nuclear factor κB (NF-kB) mediates the shift in the melatonin source from pinealocytes to immune competent cells." *International journal of molecular sciences*, 14(6), 10979–10997.

36 Rodríguez, V., Mellado, C., Alvarez, E., Diego, J. G. & Blázquez, E. (1989). "Effect of pinealectomy on liver insulin and glucagon receptor concentrations in the rat." *Journal of pineal research*, 6(1), 77–88.

37 Bonilla, E., Valero, N., Chacín-Bonilla, L. & Medina-Leendertz, S. (2004). "Melatonin and viral infections." *Journal of pineal research*, 36(2), 73–79.

38 Reiter, R. J. (1973). "Pineal control of a seasonal reproductive rhythm in male golden hamsters exposed to natural daylight and temperature." *Endocrinology*, 92(2), 423–430.

39 Cipolla-Neto, J., Amaral, F.G., Afeche, S.C., Tan, D.X. & Reiter, R. J. (2014). "Melatonin, energy metabolism, and obesity: a review." *Journal of pineal research*, 56(4), 371–381.

40 Blavatsky, H. P. (1895). *The Secret Doctrine: The Synthesis of Science, Religion and Philosophy* (Vol. 2). Theosophical Publishing Society.

41 Bible, K.J., & Various. (1996). *King James Bible*. Project Gutenberg.

42 Hall, M.P. (1938). *Occult anatomy of man*. Chicago.

43 Lumry, R., & Hershberger, M. (1978). "Status of indole photochemistry with special reference to biological applications." *Photochemistry and Photobiology*, 27(6), 819–840.

44 Angiolillo, P.J., & Vanderkooi, J.M. (1996). "Hydrogen atoms are produced when tryptophan within a protein is irradiated with ultraviolet light." *Photochemistry and photobiology*, 64(3), 492–495.

45 Borkman, R.F. & Lerman, S. (1978). "Fluorescence spectra of tryptophan residues in human and bovine lens proteins." *Experimental eye research*, 26(6), 705–713.

46 Azmitia, E.C. (2007). "Serotonin and brain: evolution, neuroplasticity, and homeostasis." *Int Rev Neurobiol*, 77, 31–56.

47 Sturgis, J.N., Olsen, J.D., Robert, B. & Hunter, C.N. (1997). "Functions of conserved tryptophan residues of the core light-harvesting complex of Rhodobacter sphaeroides." *Biochemistry*, 36(10), 2772–2778.

48 Godik, V.I., Blankenship, R E., Causgrove, T.P. & Woodbury, N. (1993). "Time-resolved tryptophan fluorescence in photosynthetic reaction centers from Rhodobacter sphaeroides." *FEBS letters*, 321(2), 229–232.

49 Vavilin, D.V., Ermakova-Gerdes, S.Y., Keilty, A.T., & Vermaas, W.F. (1999). "Tryptophan at position 181 of the D2 protein of photosystem II confers quenching of variable fluorescence of chlorophyll: Implications for the mechanism of energy-dependent quenching." *Biochemistry*, 38(44), 14690–14696.

50 Crocker, E., Eilers, M., Ahuja, S., Hornak, V., Hirshfeld, A., Sheves, M. & Smith, S. O. (2006). "Location of Trp265 in metarhodopsin II: implications for the activation mechanism of the visual receptor rhodopsin." *Journal of molecular biology*, 357(1), 163–172.

51 Chabre, M. & Breton, J. (1979). "Orientation of aromatic residues in rhodopsin. Rotation of one tryptophan upon the meta I→ meta II transition after illumination." *Photochemistry and photobiology*, 30(2), 295–299.

52 Lin, S.W., & Sakmar, T P. (1996). "Specific tryptophan UV-absorbance changes are probes of the transition of rhodopsin to its active state." *Biochemistry*, 35(34), 11149–11159.

53 Tan, D.X., Zheng, X., Kong, J., Manchester, L. C., Hardeland, R., Kim, S. J., & Reiter, R. J. (2014). "Fundamental issues related to the origin of melatonin and melatonin isomers during evolution: relation to their biological functions." *International journal of molecular sciences*, 15(9), 15858–15890.

54 Tan, D.X., Hardeland, R., Manchester, L.C., Paredes, S.D., Korkmaz, A., Sainz, R.M., & Reiter, R.J. (2010). "The changing biological roles of melatonin during evolution: from an antioxidant to signals of darkness, sexual selection and fitness." *Biological Reviews*, 85(3), 607–623.

55 Tan, D.X., Manchester, L.C., Terron, M.P., Flores, L.J., & Reiter, R.J. (2007). "One molecule, many derivatives: A never-ending interaction of melatonin with reactive oxygen and nitrogen species?" *Journal of pineal research*, 42(1), 28–42.

56 Strassman, R.J. (1995). "Human psychopharmacology of N, N-dimethyltryptamine." *Behavioural brain research*, 73(1), 121–124.

57 Strassman, R. (2000). DMT: *The spirit molecule: A doctor's revolutionary research into the biology of near-death and mystical experiences.* Inner Traditions/Bear & Co.

58 Szara, S.T. (1956). "Dimethyltryptamin: its metabolism in man; the relation of its psychotic effect to the serotonin metabolism." *Cellular and Molecular Life Sciences*, 12(11), 441–442.

59 Naranjo, P. (1979). "Hallucinogenic plant use and related indigenous belief systems in the Ecuadorian Amazon." *Journal of ethnopharmacology*, 1(2), 121–145.

60 Franzen, F., & Gross, H. (1965). "Tryptamine, N, N-dimethyltryptamine, N, N-dimethyl-5-hydroxytryptamine and 5-methoxytryptamine in human blood and urine." *Nature*, 206(4988), 1052–1052.

61 Barker, S.A., McIlhenny, E.H. & Strassman, R. (2012). "A critical review of reports of endogenous psychedelic N, N-dimethyltryptamines in humans: 1955–2010.:" *Drug testing and analysis*, 4(7–8), 617–635.

62 Barker, S.A., Monti, J.A. & Christian, S.T. (1980). "Metabolism of the hallucinogen N, N-dimethyltryptamine in rat brain homogenates." *Biochemical Pharmacology*, 29(7), 1049–1057.

63 Christian, S.T., Harrison, R., Quayle, E., Pagel, J. & Monti, J. (1977). "The in vitro identification of dimethyltryptamine (DMT) in mammalian brain and its characterization as a possible endogenous neuroregulatory agent." *Biochemical medicine*, 18(2), 164–183.

64 Sitaram, B.R., Lockett, L., Talomsin, R., Blackman, G.L. & McLeod, W. R. (1987). "In vivo metabolism of 5-methoxy-N, N-dimethyltryptamine and N, N-dimethyltryptamine in the rat." *Biochemical pharmacology*, 36(9), 1509–1512.

65 Takahashi, T., Takahashi, K., Ido, T., Yanai, K., Iwata, R., Ishiwata, K., & Nozoe, S. (1983). *11C-Labeling of Indolealkylamine Alkaloids and the Comparative Study of Their Biodistributions.*

66 Cozzi, N.V., Gopalakrishnan, A., Anderson, L L., Feih, J.T., Shulgin, A.T., Daley, P F. & Ruoho, A.E. (2009). "Dimethyltryptamine and other hallucinogenic tryptamines exhibit substrate behavior at the serotonin uptake transporter and the vesicle monoamine transporter." *Journal of neural transmission*, 116(12), 1591.

67 Frecska, E., Szabo, A., Winkelman, M.J., Luna, L.E. & McKenna, D.J. (2013). "A possibly sigma-1 receptor mediated role of dimethyltryptamine in tissue protection, regeneration, and immunity." *Journal of Neural Transmission*, 120(9), 1295–1303.

68 Szabo, A., Kovacs, A., Frecska, E. & Rajnavolgyi, E. (2014). "Psychedelic N, N-dimethyltryptamine and 5-methoxy-N, N-dimethyltryptamine modulate innate and adaptive inflammatory responses through the sigma-1 receptor of human monocyte-derived dendritic cells." *PLoS One*, 9(8), e106533.

69 Meyer, P.(2010, August, 20) *340 DMT Trip Reports*. Retrieved from http://www.serendipity.li/dmt/340_dmt_trip_reports.htm

70 Popp, F.-A. (1999). "Macroscopic Quantum Coherence", *Proceedings of an International Conference on the Boston University*, edited by Boston University and MIT, World Scientific.

71 Macchi, M.M., & Bruce, J.N. (2004). "Human pineal physiology and functional significance of melatonin." *Frontiers in neuroendocrinology*, 25(3), 177–195.

72 Tilden, A.R., Becker, M.A., Amma, L.L., Arciniega, J., & McGaw, A.K. (1997). "Melatonin production in an aerobic photosynthetic bacterium: an evolutionarily early association with darkness." *Journal of pineal research*, 22(2), 102–106.

Qi-Water Bridge as a
Bilateral Consciousness-Brain Interface

NELSON ABREU

Introduction

In this chapter, a *qi-water* bridge is proposed as a means of mediation between body (*bios*) and being or consciousness (*conscientia*), allowing for the firing of neurons. A definition of qi (chi, biofield, bioenergy) based on shared meaningful information is advanced. Additionally, the brain-being interaction is further qualified by suggesting that the consciousness communicates with each brain hemisphere in a distinct way, integrates both pathways into a seamless sense of self.

The biofield is widely regarded as a foundational concept for modeling how an extra-material being or consciousness might manifest in the material dimension. If the source of consciousness is beyond the brain, it should connect to the brain in some way in order to vitalize the body as a whole and facilitate mind-matter interactions and other psi phenomena through this medium. Yet, for a concept that is so heavily relied upon, a satisfactory definition remains lacking in the emerging science of consciousness. It is even less clear how much such a Source could form a practical interface with the material brain.

Some scholars of consciousness anomalies, as well as contemporary and ancient qi and Out of Body Experience (OBE) practitioners, have discussed a vital force or biofield as a medium for this inter-dimensional communication. This hypothetical model of brain-"parabrain" communication remains vague, however, since the nature and behavior of the biofield is poorly understood. Emerging scholarship may be converging on more detailed models for how a nonlocal consciousness may interact with the material theater, including the brain, as a potential extension or reflection of consciousness, an information system, or even a sort of technology of consciousness. In the present book, the reader can address this fundamental question by piecing together insights from physics, mathematics, engineering, neuroscience, biology, and the study of consciousness anomalies. In this chapter we seek to integrate such insights with the curious properties of water, which make it a major molecule of life. With this collection of works, a plausible mechanism emerges from various lines of study that allows for one or more beings (such as a species, like cockroaches) to attract a desired outcome (decreased vulnerability to a predator, like humans), making a favorable genetic mutation more likely (such as a newly-found distaste for bait poison, which was previously perceived as desirably sweet). One piece of the puzzle, however, remains elusive. How does a being that is not limited to the material brain communicate with it?

> Omnis cellula e cellula (every cell comes from a cell)
> — Rudolph Virchow (1858)

> Omne vivum ex vivo – Life comes from Life
> — Louis Pasteur (1864)

Water: Life's Matrix

From high-pH environments to deep-sea hydrothermal vents without the benefit of sunlight, it seems that wherever there is water, we are bound to find life. We have yet to identify a single life form that can exist in its absence. Biologists point to special properties of water that make it an excellent solvent at typical Earth temperatures, a mediator of life's delicate chemical reactions unlike any other, and it has a highly-structured local order uncommon among liquids. It is no wonder that Nobel laureate Albert von Szent-Györgyi described water as the "matrix of life," since biological processes take place with and within water (Szent-Gyorgyi, 1977).

All life uses a membrane to negotiate with its environment, filtering in materials for energy and sealing off and ejecting harmful substances such as waste products. Water can also change states within a relatively narrow range of temperatures, making it able to preserve the seeds of life in meteorite ice. Life on Earth began in the oceans. In all likelihood, without water life simply cannot be sustained: we die quickly without it.

It is even a structural component of plants, allowing flowers to hold up to face the Sun. There are also examples of water bound to proteins playing an important functional role. For example, chains of water molecules inside protein pore-like channels can act as "proton wires." (Ball, 1999).

Ulisse di Corpo and Antonella Vannini (2010) have proposed that the anomalous properties of water, in particular the hydrogen bridge, could create a mechanism for the consciousness to affect matter. (See their chapter in this book.) Through water, consciousness may bridge the gap between the subatomic and macroscopic levels, making this molecule essential for life. These unusual properties of water could have led it to be become "biophilic," a sort of "molecule of life," associated with all known forms of biological life. Through a principle of similarity, water may also have the tendency to present itself in low-entropy, structured states that make it attractive to consciousness as an order-affecting agent. Consciousness appears to be able to affect the order of a system. Water may be more sensitive to bioenergy, because it is sensitive to the consciousness as an ordering principle, which could explain why water has proven an essential ingredient for any and all forms of biological life.

Water is a dipole and acts like a magnet, with the oxygen end having a negative charge and the hydrogen end having a positive charge. These charged ends can attract other water molecules, forming strong hydrogen bonds through this intermolecular force. Conventional science states that there are three phases of water: solid, liquid and vapor, which can exist simultaneously. However, recent studies by Professor Gerald Pollack, from the University of Washington, point to a so-called fourth phase of water (Pollack, 2013).

This fourth phase, also called the Exclusion Zone (EZ) water because it excludes impurities, occurs next to "water-loving" surfaces (hydrophilic materials). It extends almost everywhere in nature, including the human body, and it can be increased through electric charge or exposure to light (electromagnetic energy), whether in the form of visible, ultra-violet or, especially, infrared light. Water is unusually viscous and dense, but EZ water is even more so, as well as being more alkaline and negatively charged. Pollack provides evidence that water molecules in the cell are very near to one or another hydrophilic surface and therefore more ordered.

The body consists of over 99 percent water, but according to Pollack and others, the water in cells is not regular water, but rather this highly structured water with special properties. This fourth phase of water, structured as H_3O_2 rather than H_2O, may provide conditions for a non-material process to employ the biofield as a means of creating meaningful informational patterns to affect material reality via the brain. Whatever makes water so essential for biological life, it may be what makes it a perfect bridge molecule for consciousness to act upon the physical body. (Abreu and Tordoledo, 2017) Could it behave in a similar fashion when exposed to chi?

The idea that cellular processes may involve coherent mechanisms is not new. Schrödinger argued that "negative entropy" was at play and that such ordering mechanisms are exactly what would be needed to explain life (Schrodinger, 1944). Nobel laureate Albert Szent-Gyorgyi, commonly regarded as the father of modern biochemistry, claimed that water's role in biology was central: "Life is water dancing to the tune of solids" (Szent-Gyorgyi,1971). The most plentiful molecule in biological life, very likely "the molecule of life," has been largely forgotten today, but interest in the field is being revived by some.

Chemist and physicist Philip Ball, author of *Life's Matrix: A Biography of Water*, is among the scientists who recognize that learning more about the structure of cell water could explain more about why it is so important for biology. In an interview with NASA, he remarked that "very recently there have been several reports that water very near hydrophobic surfaces is vapor-like: that such surfaces are relatively 'dry'. This would have important implications for things like the hydrophobic attraction, as well as for the nucleation of gas bubbles at surfaces, which has been proposed as an explanation for the mysteriously long-ranged attraction between hydrophobic surfaces" (Ball, 1999).

Gilbert Ling was one of the earliest proponents of the role that structured, bound, bulk, or low-entropy water could have in cell physiology, rather than the more commonly accepted diffusion, pump or membrane theories (MT). According to Ling, the cell is a holistic system because water and potassium ions are adsorbed to a matrix of cell proteins with natively unfolded conformation, which strongly binds, orients and polarizes water (Ling, 1977).

Since water, making up the majority of the cell, may be ordered, bound, limited in motion in a fourth phase that is gel- or colloid-like, the entropy of the cell is lower. The bound state of ions and the network of interconnected protein molecules (cell matrix) further reduces the entropy. All these features are inherent to the cell during the resting state (R-state). During the protoplasmic phase transition from the R-state to an active state (A-state), such as neuron firing (action potential), the complex is transformed and water and K+ become free and the entropy of the system increases (Matveev, 2012). The released energy produces heat, volume change, and is utilized for biological work.

It is remarkable how much we still do not know about the brain. How a cell, like a neuron, truly works is only part of the mystery. A much more elusive matter is why it fires. The focus is usually on neural activity in response to external stimuli. Most things we care about, however, like memory, emotion, drive, can occur with no external stimulus and no overt output that can be measured. Brain activity is not limited to the kind that can be measured, for its networked and massive nature is not possible to fully capture and it is not limited to neurons. The brain also has another kind of cell called glia. Originally thought to be a mere structural support for neurons, it has been discovered that these cells can also communicate with neurons and other glia cells. Whereas neurons communicate electrically and chemically, glia only communicate chemically and appear involved in every aspect of neural function. "Glial cells maintain the brain's environment, regulate synapses and neurotransmitters, respond to injuries, and in certain cases can even become neurons" (Jabr, 2012).

We can see the plausibility that consciousness, being able to introduce order into otherwise high entropic, random or stochastic systems, as shown in research conducted at PEAR and elsewhere, could trigger an action potential in a complementary way by affecting any system in a delicate dynamic equilibrium (Abreu and Tordoledo, 2017). If the neuron is a meta-stable entity, thanks to the plentiful occurrence of structured water in our brain, consciousness may be able to induce significant change with a small influence. In this case, however, rather than introducing order into a stochastic system, it is possible that consciousness could tip the scales on a meta-stable ordered system, causing action potentials. Even if the signal is weak compared to existing electrical noise in the brain, coherently vibrating molecules might sum up their response for a substantial effect that could trigger neural activation.

More research is required to continue expanding the body of knowledge on the consciousness-brain link, to incorporate theories such as syntropy and discoveries such as the fourth phase of water. We can consider the possibility that bioenergy is related to similarity and that bioenergy and water play an important role in the connection between Mind and matter.

> As scientists attempt to understand a living system, they move down from dimension to dimension, from one level of complexity to the next lower level. I followed this course in my own studies. I went from anatomy to the study of tissues, then to electron microscopy and chemistry, and finally to quantum mechanics. This downward journey through the scale of dimensions has its irony, for in my search for the secret of life, I ended up with atoms and electrons, which have no life at all. Somewhere along the line life has run out through my fingers. So, in my old age, I am now retracing my steps, trying to fight my way back.
>
> — Szent-Györgyi A. (1972)

Qi as a Measure of Shared Meaningful Information

Bioenergy is often invoked to provide a bridge between a non-material consciousness and the material brain. What is bioenergy, after all? Is it a form of energy? A kind of matter or substance? A sensation or feeling? We suggest it is a principle related to the state and nature of consciousness itself, which reflects increasing similarity, cooperation, resonance, harmony, coherence, order, or syntropy. As previously discussed, the curious properties of water could make it particularly conducive for the consciousness to express itself through this link. Bioenergy has been referenced by nearly all cultures around the world and has been discussed by thinkers throughout history. Qi is often discussed as some form of energy and is typically translated as "life-force energy" or "bioenergy." An adequate model of Qi based on physical energy concepts alone does not account for its objective and subjective qualities, possibly because it is not the most suitable form to think of it. At the 2015 International Congress on Consciousness, an alternative model of Qi was proposed to eliminate some of the shortcomings of the energy model. It related bioenergy to "cooperation," as the coordination of a distributed group of systems toward the achievement of a desired outcome. (Anderson and Abreu)

Here we re-frame Qi as *similarity* of desired outcome, values or rapport. This outcome can be the state of cooperation itself. Interestingly, cooperation in this sense, requires consciousness; and consciousness is the field in which a goal or desired outcome exists. Thus, the notion of cooperation bridges the gap between the intangible subjective realm of consciousness, mind, and will and the tangible objective physical realm of actions and measurables. Cooperation and Qi both exist where consciousness interacts with environments like physical systems. Thus, information units, such as *terabytes* of shared meaningful information, could be a more useful unit for chi than energy units like Joule or BTU. The greater the meaning and volume of shared information, the greater the similarity, rapport and "energetic attraction" between the entities.

Bioenergy, then, might be seen as a manifestation of cooperation, similarity or resonance with the consciousness itself (health, balance, harmony with its nature as an evolving, changing, complexifying, creative, ordering principle); with the cosmos (cosmoethics, cosmic union); with other beings, and with extensions such as the nonlocal vehicles of manifestation and technology. This harmony with our own nature (consciential health) can be interpreted as vitality, health and happiness (well-being).

The "transmittal" of this harmony does not behave like a normal field that decays with distance, any more than pixels representing objects on a computer game. Rather, it resembles more the workings of data within a computer or computer network. When bioenergy reaches a resonant state, it produces tangible vibrations and other sensations, yet Qi is not exclusively a substance or biomatter. Qi also seems to intensify or be shared more whenever there is a reduced sense of separation (sympathetic assimilation, friendship, love, cosmic unity). Rapport arises from a high similarity with others, like overlapping semiotics Venn diagrams showing similarity or overlap of meaningful informational content (Fig. 1). Information, objects, ideas pertaining to oneself, including one's past or potential future states would share a significant overlap with the present set of information due to self-similarity. When we sense our own Qi, in our body, environment, belongings, or familiar places, we may be experiencing this self-similarity.

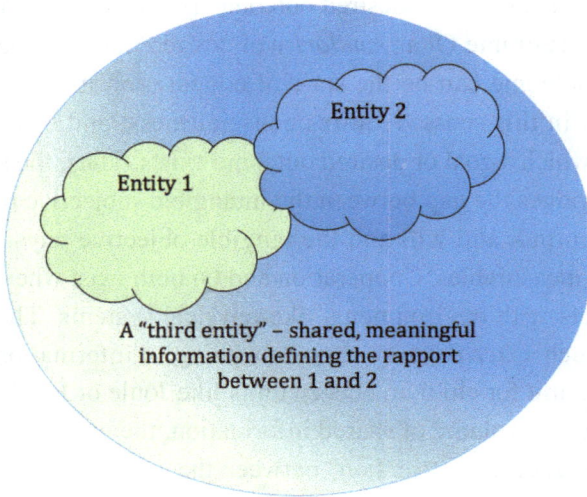

Figure 1 – Qi could be related to the integration of information shared between entities and the meaning attributed to said information

Consciousness appears to be able to introduce similarity, cooperation, harmony or order (syntropy) into localized portions of the universe (sand into sand castles, matter-energy into biological systems, uncoordinated chi into a vibrational state, individuals into cooperative friends or synergistic groups). It also responds to the degree of similarity or difference with its environment or other beings. Resonance (or cooperation) can be seen as a metaphor for similarity, coherence or convergence in attention, purpose, intention, meaning, emotional states, or other aspects among individuals. Resonance, then, with this metaphorically increased order in the subjective world, and the resulting reduction in informational, morphic, semiotic or "thosenic entropy," would be related to an increase in syntropy in the realm of consciousness.

The work of the PEAR laboratory and elsewhere suggests this increased "resonance" in the subjective sphere has a comparable effect in the physical realm, as reflected in increased order in random number generators and other stochastic physical devices used in experiments. Resonance, therefore, may be seen as an expression for bioenergy. A practical example could might be when individuals are in a more sympathetic state and their life energies may temporarily merge (auric coupling, sympathetic assimilation). How appropriate, then, that when our bioenergy fields (energy body or energosoma) are at their peak level of health or order, they seem to reach what is perceived as a harmonic, resonant state known as vibrational state.

Curiously, ongoing fMRI research on the vibrational state described in Wagner Alegretti's chapter in this book demonstrates anomalies consisting of detection of "brain activity" in the space outside the brain, or even within certain objects placed in the fMRI machine. In other words, it appears that the vibrational state (ordered chi) and exteriorization of chi (shared Qi) change sub-atomic properties of matter in and around certain targets such as the brain, an egg, or even the phantom fluid used in fMRI studies.

Note that water is one of the common elements present in all three, reinforcing the notion that water is highly responsive to qi effects due to its properties. Since the fMRI detects coherence in the spin of protons, this effect could be interpreted as consciousness instilling a greater degree of order (reducing entropy or increasing syntropy) in the physical realm, which we interpret as bioenergetic states. Similarly, individuals connecting emotionally with a machine, or a large number of individuals in emotional resonance with one another, seems to reduce the randomness of stochastic systems.

While the suggestions presented in this paper remain speculative and require additional vetting, the main purpose remains to invite consciousness scholars to collaborate to reach a more complete model of the purported link between extra-material consciousness and the brain. A concept that is so commonly evoked in modeling complementary alternative medicine, psi phenomena, energy mindfulness practices and other related subjects should be better understood.

A shift from thinking about gradual selection of localized random changes to sudden genome restructuring by sensory network-influenced cell systems is a major conceptual change. It replaces the "invisible hands" of geological time and natural selection with cognitive networks and cellular functions for self-modification. The emphasis is systemic rather than atomistic and information-based rather than stochastic

> — James Shapiro, Evolution: A View from the 21st Century. Upper Saddle River, NJ: FT Press, 2011

Hemispheric Asymmetry

As we have moved toward an information- and meaning-based approach to the connection between consciousness, physical reality, and the brain, we can continue to look to computer science for inspiration. Parts of a computer system deal with input and output, interacting with the user and sources of power and data. Other components operate more internally, with little direct interaction with the outside world. Could a consciousness-centric paradigm have explanatory power for why the brain has two hemispheres? One sub-system of brain circuits, not entirely confined to, but largely active in one hemisphere (often the left), could be considered predominantly environment-facing. The other, could be more extra-material or consciousness-facing.

Evidence suggests that the left hemisphere tends to generally concern itself with details or the "local picture," whereas the right hemisphere's activity seems tilted towards the "global picture." Initial support for this framework is drawn from hints such as physiological differences and research that includes "split-brain" patients, brain imaging during visual tasks with temporarily-disabled hemispheres, or with alternating emphasis of either detailed or global focus (Christie, J. et al, 2012).

It would be misleading to refer to anyone as either left- or right-brained without excessive simplification. While some functions like language tend to be concentrated in the left hemisphere, there is a percentage of the population that has it on the right hemisphere. These differences do not correspond 100% to handedness, either. It is clear, however, that brain function is not symmetrical. Old ideas about humans using only a small percentage of their brains also seem misleading today. Many activities seem to activate various parts of the brain simultaneously, working like several members of an orchestra. The "self" is not located in a particular area of the brain, as far as neuroscience can tell, but rather seems to be distributed, as if emerging from the brain's "symphony" of similarly-timed events. The brain is also incredibly adaptable and can change substantially over time and in response to surgery or injury.

We can consider a tentative framework to explicate why the brain has two hemispheres with its different styles, drawing from computer design principles. Initial support for such a framework is drawn from hints such as physiological differences and research that includes "split-brain" patients, brain imaging during visual tasks with temporarily-disabled hemispheres, or with alternating emphasis of either detailed or global focus.

The activity of two contiguous but separate hemispheres does not result in two separate consciousnesses, thus undermining the idea that consciousness arises from the brain. There are other possible, sub-cortical pathways, that can provide some connectivity between hemispheres, but studies suggest that we can have "two brains" with distinct attitudinal frameworks. When they are separated, one might understand the words of the joke (typically the left hemisphere through the right ear), while the other hemisphere would cause us to laugh, but not quite sure why. One hemisphere will attempt to create a narrative to understand the facts, the other appears to process meaning or relevance and what is often called "theory of mind" (predicting or sensing what others might think or feel). As another example, when we listen to a song, one hemisphere may focus on the words and the other on the music.

Studies on split-brain patients and related investigations that examine disruption in inter-hemispheric communication or simultaneity of bicameral function (such as cases of stroke) suggest something new for a consciential paradigm: the consciousness appears to be connected to both hemispheres, but in different ways. Part of the brain, typically regions in the left hemisphere, are more objectively oriented and are more involved with language processing, speech, and dealing with details. Other regions, typically in the right hemisphere, are more subjectively oriented, and deal with the big picture and meaning.

For instance, when split brain patients are provided conflicting information to each hemisphere, their arms may struggle with each other to point at the "correct" picture. They are still able to process language, but in more symbolic, visual ways, unable to say the word they read, but able to point to the corresponding picture. However, both hemispheres communicate with a unified mind that is able to retain a singular sense of self.

Split-brain research pioneer Michael Gazzaniga developed an "interpreter theory" to explain why people have a unified sense of self and mental life, even after undergoing a corpus callosotomy. (Gassaniga, 1999) They might become aware that they have two distinct ways to experience the world, they may even experience an "alien hand" type of syndrome, but their sense of self is not split. The theory was developed based on observations of these patients that indicated that when asked to explain in words, which uses the left hemisphere, an action that had been directed to and carried out only by the right one, the left hemisphere of split-brain patients made up a post-hoc answer that fit the situation.

In one case, Gazzaniga flashed the word "smile" to a patient's right hemisphere and the word "face" to the left hemisphere, and asked the patient to draw what he'd seen. His right hand drew a smiling face and made up a story about why the face was smiling. The left-brain interpreter seeks to construct narratives that help to make sense of the world and the barrage of information it provides. In this case, it did so with information from the right hemisphere that the left brain appeared unaware of, but that the unified mind was able to weave together.

For most people both lateral facets are integrated and in constant communication. However, depending on which is predominant, physical awareness and behavior can come through differently. This has been further demonstrated by the "God helmet" experiments at Laurentian University, where creative flair was show to be enhanced and reading ability altered when the left hemisphere was subjected to electromagnetic "jamming" (Persinger, 2002).

It is also noteworthy that studies of certain meditation techniques and the vibrational state indicate an increase in cross-hemisphere communication (increase in EEG gamma brain waves and fMRI activation seen throughout both hemispheres). One "side"—or more accurately, one distributed network set—of the brain is more concerned with details, while the other is dedicated to the whole (McGilchrist, 2009). I propose that one part of the brain is mostly concerned with interfacing with the physical "outside world" through language, calculation, navigation, action, and conscious awareness. The other is interfacing more with the extraphysical or microcosm, biased toward meaning, symbolism, intention, and the sub-conscious. Such activities as the Voluntary Energetic Longitudinal Oscillation (VELO) and certain types of mindfulness practices could thus be seen as increasing the communication between the "big picture" consciousness itself (big M "Mind") and the "detail-oriented" physical counterpart (small "m" mind). Such practices appear to be integrating the different "polarities" or attributes of the consciousness (e.g. intuition and reason). The predominance of gamma waves in EEG studies of such practices underscores this possibility, as they are associated with enhanced communication between distant parts of the brain, across hemispheres.

Conclusion

In the foregoing chapter, we have attempted to integrate various lines of research and thought to arrive at a definition of chi that incorporates its numerous facets. A framework based on shared or similar meaningful information was put forth to explain qi, combined with the evidence that shows that consciousness is capable of affecting the degree of order in a system. We have argued that the unique properties of water could make it ideally suited to compose a bridging agent between the body and the consciousness via qi. Some researchers have described a highly structured but only marginally stable state of water—sometimes described as a colloidal fourth phase between liquid and solid—present in the membranes of cells such as neurons. The delicate equilibrium state could be affected by aforementioned ability to affect the order of physical systems, triggering action potential involved in electrical and chemical signaling in the nervous system. This type of Qi-Water Bridge for Consciousness-Brain interaction could explain, in part, why water has been a ubiquitous ingredient of living systems.

Finally, the interaction of the consciousness with the brain can be further qualified by exploring functional asymmetries of the brain, particularly expressed in split-brain patients. Studies point to one hemisphere acting largely as an external detail-oriented, emissary of the right hemisphere, which tends to focus on the big picture. The integration of these two biological mind frames, however in tension, still result in a unitary experience, even in split brain patients.

While the suggestions presented here remain speculative and require additional vetting, our main purpose remains to invite consciousness scholars to collaborate to reach a more complete model of the purported link between extra-material consciousness and the physical brain. A concept that is so commonly evoked in modeling complementary alternative medicine, psi phenomena, energy mindfulness practices and other related subjects should be better understood.

Endnotes

Abreu, Nelson & Machin, Pedro (2006). *Consciential Asymmetry: Toward a Non-Reductionistic Framework and Ontology of Brain Function Laterality*. Presentation at the 25th Annual Meeting of the Society for Scientific Exploration, Orem, Utah, USA, June 8–10.

Abreu, Nelson, Madurell, Alexandre & Perego, Lucilla (2013). *The Consciential Paradigm: A consciousness-centered framework for expanding the study of reality through bioenergy, OBE, and allied phenomena*. Presentation at the 1st International Conference "Life Energy, Syntropy and Resonance," Viterbo, Italy, 1-4 August.

Abreu, Nelson & Tordoledo, Joel (2017). *The Inter-dimensional Link Between Physical and Non-physical Consciousness*. Presentation at the International Congress on Consciousness, Miami, Florida, USA.

Ball, P. (1999). H2O: *A Biography of Water*, Weidenfeld & Nicolson, London; Brogaard, Berit (2012). "Split Brains." Psychology Today, 6 November, website, psychologytoday.com/blog/the-superhuman-mind/201211/split-brains, accessed 5 February 2015

Christie, J. et al (2012). Global versus local processing: seeing the left side of the forest and the right side of the trees. Frontiers in Human Neuroscience, 6, 28, Website: http://doi.org/10.3389/fnhum.2012.00028, accessed 8 September, 2017.

Di Corpo, Ulisse & Vannini, Antonella (2010). "Syntropy and Water." *Syntropy*, pp.82–87, website, www.lifeenergyscience.it/english/2010-eng-1-3.pdf, accessed 5 February 2015.

Dunne, Brenda & Jahn, Robert (2005). "Consciousness, Information, and Living Systems." *Cellular and Molecular Biology*, 51, R. Wegmann, Noisy-le-Grand, France, pp.703–714.

Dunne, Brenda & Jahn, Robert (2015). "Consciousness and the Nature of Life." *Proceedings of the 1st Congress on Consciousness. Journal of Consciousness*, Vol. 18, Issue 59.

Eagleman, David (2007). "10 Unsolved Mysteries of The Brain." Discover Magazine, 31 July, website, discovermagazine.com/2007/aug/unsolved-brain-mysteries, accessed 5 February 2015.

Fukuyama, Hidenao (2010). *Water: The Forgotten Biological Molecule*, Taylor & Francis Group, Boca Raton, Florida, USA.

Gazzaniga, Michael (1999). "The Interpreter Within: The Glue of Conscious Experience." *Cerebrum*, The Dana Foundation. http://www.dana.org/Cerebrum/Default.aspx?id=39343

Gerschenfeld, Ana (2014). "Primeira observação da estrutura da água líquida a… 46 graus negativos." Publico, 18 June, website, www.publico.pt/2014/06/18/ciencia/noticia/primeira-observacao-da-estrutura-da-agua-liquida-a-46-graus-negativos-1659544, accessed 5 February 2015.

Jabr, Ferris (2012). Know Your Neurons: Meet the Glia. Scientific American Blogs: Brainwaves. 18 May. Website: https://blogs.scientificamerican.com/brainwaves/know-your-neurons-meet-the-glia/, accessed 4 September, 2017.

Jaeken, Laurent & Matveev Vladimir (2012). "Coherent Behavior and the Bound State of Water and K+ Imply Another Model of Bioenergetics: Negative Entropy Instead of High-energy Bonds." *The Open Biochemistry Journal*, USA, 6, pp.139–159.

Ling, Gilbert (1977). "The physical state of water and ions in living cells and a new theory of the energization of biological work performance by ATP." Molecular and Cellular Biochemistry, Volume 15, May, Issue 3, pp 159–172.

McGilchrist, Iain (2009). *The Master and His Emissary: The Divided Brain and the Making of the Western World*. USA: Yale University Press.

Nichols, Wallace (2014). *Blue mind: How Water Makes You Happier, More Connected and Better at What You Do*, Little Brown and Company, USA.

Persinger, Michael A; Healey, Faye (2002). "Experimental facilitation of the sensed presence: possible intercalation between the hemispheres induced by complex magnetic fields." *The Journal of Nervous and Mental Disease*. 190 (8): 533–41

Pollack, Gerald (2013). *The Fourth Phase of Water*, Ebner and Sons, Seattle, Washington, USA.

Schrodinger, E. (1944). *What is Life*. Cambridge University Press.

Szent-Gyorgyi, A. (1971). "Biology and Pathology of Water." *Perspectives in Biology and Medicine*, Volume 14, Number 2, Winter, pp. 239–249

Szent-Gyorgyi, A. (1972) "What is life?" *Physical basis of life*. pp. 5, Del Mar, CA: CRM Books

Szent-Gyorgyi, A. (1977). "Drive in Living Matter to Perfect Itself." *Synthesis* 1, Vol. 1, No. 1, 14–26.

Tamagawa, Hirohisa et al. (2016). *Ling's Adsorption Theory as a Mechanism of Membrane Potential Generation Observed in Both Living and Nonliving Systems*, Membranes, 6, 11, Multidisciplinary Digital Publishing Institute, Basel, Switzerland, January.

Wolman, David (2012). "The split brain: A tale of two halves," *Nature*, 14 March, website, www.nature.com/news/the-split-brain-a-tale-of-two-halves-1.10213, accessed 5 February 2015.

The Physical Nature of the Biological Signal: From High Dilutions to Digital Biology

YOLÈNE THOMAS

Abstract

The memory of water was a radical idea that arose in the laboratory of Jacques Benveniste in the late 1980s. Over thirty-five years have passed and yet the often angry debate on its merits continues, despite the increasing number of scientists who have reported confirmation of the basic results. A parallel can be drawn between this debate on the memory of water, which presumes that the action of molecules is mediated by an electromagnetic phenomenon, and the often acrimonious debate on the transmission of nerve influxes via synaptic transfer of specific molecules, neurotransmitters. The latter debate began in 1921 with the first experiments by Loewi and was still active in 1949, 28 years later. A strong reluctance to accept research that questions basic aspects of long-accepted biochemical paradigms is to be expected. In this paper, we will provide a brief summary of experiments relating to the memory of water: the earlier work on high dilutions (HD) and then the experiments, which followed and continue today, on digital biology.

UHD 1994: The Early History of High Dilutions Experiments

Despite the difficulties after the *Nature* fracas in 1988, Benveniste and his team, of which I was member, pursued research to understand the physical nature of the biological signal in HD. In UHD 1994, Benveniste presented several studies on the biological effects of agitated highly diluted substances: 1. on cell-lines, 2. on isolated guinea-pig heart (Langendorff), and 3. *in vivo* in a mouse model.

1. In the wake of heavy metal poisoning, serious disorders, either inflammatory or strictly immunological, occur. We studied the effects of one such heavy metal, cadmium (Cd), to determine its potential effects at very low doses. When human cell-lines were cultured in the presence of 5 to 10 μ M Cd, a high mortality rate was observed. However, when they were pretreated with ponderal, though non-toxic, doses or with HD of Cd (dilution log 16-25 or 26-35) for several days, a significant modulation of cellular activation and growth was observed, either directly, before the addition of toxic concentrations of Cd, or after it.

2. Isolated guinea pig or rat hearts were perfused at constant pressure in a Langendorff system with highly diluted vasoactive amines. Acetylcholine (ACh), histamine (H) or water (W) was injected via a catheter just above the aorta. Variation in coronary flow (CF) was measured every minute for 30 minutes. At the same time, other mechanical parameters (min. and max. tension, heart rate) were also recorded. The percent (%) increase in CF was calculated as follows: [1 −(CF maximal value / CF time 0 value)] x 100. A significant time-dependent modification ($p < 0.001$) of the guinea pig heart CF was induced by histamine dilutions (log 31-41) but not by the diluted / agitated buffer (diluted histamine vs diluted buffer, $p > 0.05$).

3. A collaboration with an external team of physicists (Lab. Magnétisme C.N.R.S. Paris) showed in 24 blind experiments that the activity of HD histamine was abolished either by heating (70°C, 30 minutes) or exposure to a magnetic field (50 Hz, 15 x 10-3 T, 15 minutes) which had no comparable effect on the genuine molecule. The action of HD of silica, a substance that, in ponderal doses, is cytotoxic for macrophages, was studied *in vivo* to determine its impact on the synthesis, of paf-acether, an ether-lipid mediator of inflammation and its inactive precursor, lysopaf-aceter by mouse peritoneal macrophages. The macrophages from silica-treated mice were stimulated *in vitro* by zymosan. Paf-aceter production was amplified from 44.2 to 67.5%, in HD

experiments, as compared to control mice. These differences were highly significant in all experiments ($p < 0.01$ to $p < 0,05$). There was no effect on precursor lysopaf-acether synthesis suggesting a cellular *in vitro* effect of HD of silica.

Possible mechanisms for the transmission of information from the molecular mother substance were discussed, including intermolecular communication by oscillating electromagnetic fields (EMF) and perimolecular coherent water separated from the substance molecule during the process of agitation (Del Guidice 1988). Together, these considerations prompted exploratory research that led to the speculation that molecules can communicate with each other, exchanging information without being in physical contact, and that at least some biological functions, which can be mimicked by certain energetic modes characteristic of a given molecule's biological signal, might be transmissible by EM means. Furthermore, it is worth pointing out that a growing number of observations suggest the susceptibility of biological systems or water to electric and low-frequency EMF. In addition, what is suggested from the literature is a possible role for EMF regarding cell communication (Albrecht-Buehler 1992, 2005; Trushin 2003; Ninham 2005; Ben Jacob 2004; Vallee 2005; Fels 2009; Cifra 2011; Chaban 2013)

1994-2014: From High Dilutions To Digital Biology

Between 1992 and 1995, the transfer of specific molecular signals to sensitive biological systems was achieved using an amplifier and electromagnetic coils. In 1995, a more sophisticated procedure was established to record, digitize and replay these signals using a multimedia computer. Briefly, the process is to capture the EM signal from a biologically active solution using a transducer and a computer with a sound card. The digital signals are stored. The signal is then amplified and 'played back' either directly to cells, organs or indirectly to water placed within a solenoid coil (Figure 1). Of note, the order of the conditions and their repetitions is always randomized and blinded. For ease in the discussion, the terminology d-X refers to the digital EMF signal from the molecules.

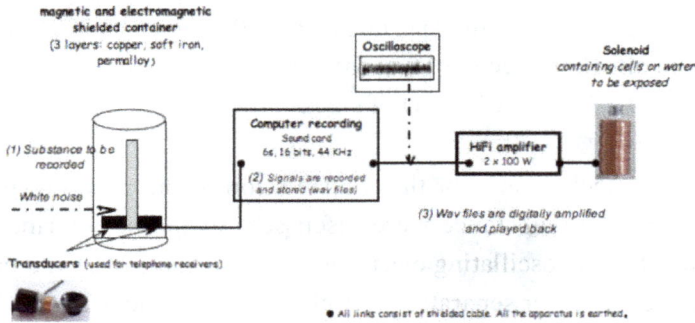

Fig. 1. Schematic drawing of the computer-recorded signals: capture, storage and replay

•Shielded cylindrical chamber: composed of three superposed layers: copper, soft iron, permalloy, made from sheets 1 mm thick. The chamber has an internal diameter of 65 mm, and a height of 100 mm. A shielded lid closes the chamber.
• Transducers: coil of copper wire, impedance 300 Ohms, internal diameter 6 mm, external diameter 16 mm, length 6 mm, usually used for telephone receivers.
• Multimedia computer (Windows OS) equipped with a sound card (5KHz to 44 KHz in linear steps).
• HiFi amplifier 2x100 watts with an "in" socket, an "out" socket to the speakers, a power switch and a potentiometer. Pass band from 10 Hz to 20 kHz, gain 1 to 10, input sensitivity +/- V.
• Solenoid coil: conventionally wound copper wire coil with the following characteristics: internal diameter 50 mm, length 80 mm, R=3.6 ohms, 3 layers of 112 turns of copper wire, field on the axis to the center 44 10^{-4} T/A, and on the edge 25 10^{-4} T/A.
All links consist of shielded cable. All the apparatus is earthed

We shall present here only three salient biological models: (1) Isolated guinea-pig heart (Langendorff); (2) Human neutrophil activation; (3) Inhibition of blood coagulation. Further details of these models have been previously described (Thomas 2006)

1. The first biological system, measurement of CF in isolated perfused guinea-pig hearts used to detect the HD effect, was also used to detect digital files endowed with biological activity. In typical experiments, the effect of digital EMF signals of acetylcholine (d-ACh) and histamine (d-H) was investigated. Digital EMF signals of water (d-W) and ACh or H, similarly were applied as negative

and positive controls respectively. The procedure and the results of consecutive blind experiments performed between November 21, 1997 and April 14, 1998 indicate that d-Ach (mean ± SD [n expts]: 19.5 ±7.4 [21]), 1 uM Ach (26.6 ±8.3 [16]), d-H (14.3 ±2.5 [14]) and 1 uM H (21.1 ±8.4 [5]) increase CF compared to d-W (4.6 ± 2.1 [28]). The two comparisons d-ACh vs d-W and d-H vs. d-W are both significant (p ‹ 0.05, Student's t test for unpaired variables). Interestingly, atropine, an ACh inhibitor, inhibited both the effects of the ACh and d-ACh but not those of H and d-H. Mepyramine, an H1 receptor blocker, inhibited both H and d-H but not ACh and d-Ach.

2. In another *in vitro* model, we investigated whether molecular signals associated with phorbol-myristate acetate (PMA) could be transmitted by physical means to human neutrophils to modulate reactive oxygen metabolite (ROM) production. Neutrophils were isolated from consenting healthy donors. PMA or vehicle were recorded, and then stored. Wav files were digitally amplified and PMA or vehicle signals were replayed for 15 minutes to neutrophils. Exposing cells to d-PMA resulted in an OD increase of 37 ± 4 % (mean ± S.E.M, 40 transmissions) compared to unexposed cells. By contrast, exposing cells to d-vehicle resulted in a 4.1 ± 1.8 % change. In the absence of cells, either d-PMA or d-vehicle was without effect on cytochrome c reduction. Furthermore, ROM was not induced when 4 a-phorbol 12,13-didecanoate (PDD), an inactive PMA analogue, was transmitted in the same manner. Data from 12 independent experiments indicate that PMA transmission (42±8 %) was essentially suppressed when a) the amplifier was turned off (-1.8±1.4 %) and b) when either the PMA solution or the neutrophils were shielded with Mu-metal (-4.3±2.7 %,). The statistical significance of the experiments was analyzed using the Student's t-test. Percent transmission was computed for each set of cells (cells exposed to d-PMA, d-vehicle, d-PDD and d-PMA amplifier off). Differences between cells exposed to d-PMA and other experimental groups were calculated at 60 minutes. d-PMA cells were associated with a 33.6 ± 3.4 % OD increase, in contrast

to 2.3 ± 1.3 % (n = 58 transmissions, p < 10 -3) for d-vehicle, d-PDD and d-PMA amplifier off (Thomas 2000).

3. The third model is the inhibition of fibrinogen coagulation by a Direct Thrombin Inhibitor (DTI) such as Melagatran. The hypothesis tested was whether the reaction rate for coagulation between thrombin and fibrinogen could be modulated by d-DTI. d-W and DTI (1uM) were used as negative and positive controls respectively. Coagulation was assessed by spectrophotometry at OD 620. Percent (%) inhibition coagulation was calculated as follows: [1 – (OD620 DTI / OD620 W)] x 100. The results of twenty-two consecutive blind experiments performed between April 16 and June 26, 2005 indicated that in most of the experiments d-DTI (mean ± SD [n expts] : 36.00 ±15.36 [22]) prolonged the clotting compared to d-W (0.09 ± 0.29 [22]) (although to a lesser extent than 1 uM DTI (70.62 ±3.42 [8]). The comparison d-DTI versus d-W was highly significant (p= 3.7 e-10, Student's t test for unpaired variables).

All together, these results suggested that at least some biologically active molecules emit signals in the form of EM radiation of less than 44 kHz that can be recorded, digitized and replayed directly to cells or to water in a manner that seems specific to the source molecules. Attempts to replicate these data in other laboratories yielded mixed results. In some cases, certain individuals consistently got digital effects and other individuals got no effects or perhaps blocked those effects (particularly when handling a tube containing informed water). Despite the precautions taken to shield the information transfer equipment from magnetic or electromagnetic pollution, very little concern has been given to possible subtle human operator effects (Dunne 2005). We dealt with this problem in some of our own studies and also in the course of one independent replication. (Jonas 2006)

Examples of Independent Experimental Work Reported by Other Groups

Distant signaling processes are not limited to interactions at the cell but also at the level of the whole organism. Between 1990 and 1994, Endler et al pioneered a series of *in vivo* experiments in which they studied the thyroxine-controlled morphogenesis regulation of the amphibian Rana temporaria from the 2- to the 4-legged stage in basins. These studies were carried out in several independent laboratories. They observed animals that were treated with HD of thyroxine (10-30M dilution) and added directly into the basin water metamorphosed more slowly than the control group, i.e. the effect of HD thyroxine was opposed to the usual physiological effect of molecular thyroxine. Other experiments showed that molecular thyroxine (1mM) or HD of thyroxine (10-30M) can be transferred via electronic circuit using water as target for the transmitted signal. Additional investigations resulted in the same effect when a sealed glass vial containing HD of thyroxine (10-30M) was simply hung into the basin water (Endler 1995, 1998). In this regard, it would be of interest to determine the interaction wavelength using vials with different thickness and material.

Since 2005, Luc Montagnier, HIV Nobelist, has been conducting experiments that would seem to confirm Benveniste's original observations. He is dealing with the detection of signals from several microorganisms, such as mycoplasma, HIV and bacteria, all derived from human pathologies. For instance, he observed that certain bacteria and/or the DNA extracted from the bacterial suspension, after being filtered and highly diluted, emitted EM waves of low frequency. The detection of the EM signals produced was analyzed using the same device as previously described (see Fig.1 on page 156).

Interestingly, this medium loses its specific EM signal when it comes into close contact with an "individual" infected by the same microorganism. A second paper shows that it is indeed possible to detect the presence of HIV DNA even when the RNA of the virus has disappeared from the blood of people with HIV who are undergoing antiviral therapy (Montagnier 2009). In a more recent paper (2011), using a polymerase chain reaction (PCR), Montagnier claimed that the DNA sequence itself could be reconstituted from the EM signal. That genes have electromagnetic representation was asserted back in 1992 by Russian scientists. This effect was called the "DNA Phantom Effect" (Gariaev et al, 1992, 2011). Still, the question remains how water can store and transmit EM information allowing a DNA sequence to be reproduced without a template.

Other emerging data are from a US group based first in La Jolla, CA, then in Seattle, WA. A company called Nativis, formerly WavBank, was founded in 2002. Since then they have conducted novel research programs and expanded the original technology that captures the unique photon field (signal) of active pharmaceutical ingredients or drugs into a series of potential industrial applications. They can improve molecular signal recording by using both magnetic and electromagnetic shielding coupled to a superconducting quantum interference device. The system records a time-series signal for a compound; the waveform is processed and optimized to identify LF peaks that are characteristic of the molecule being interrogated (Molecular Data Interrogation System). The optimized signal is played back for various periods of time to sensitive biological systems, including digital herbicides and plant growth regulators, as well as pharmaceutical compounds such as Taxol, a prototype for a class of anticancer drugs. Other experiments deal with signals derived from a therapeutic oligonucleotide, such as PCSK9 antisense RNA that lowers LDL but not HDL cholesterol.

More recently, Nativis has developed a medical device, the "Voyager" system in molecular listening, and drug delivery systems for clinical and future therapeutic treatments (Butters et al, 2014). A feasibility study of the Voyager System in patients with recurrent glioblastoma multiforme was started in December 2014. It has also being tested in veterinary applications (conf. 243rd ACS meeting, 2014).

Concluding Remarks

If these new experimental observations are validated, they will confirm and extend the Benveniste group's original findings and add other valuable pieces to the puzzle. For example: How can water represent some biologically relevant aspects of molecules? For what kind of molecules does this representation exist? What are the roles of mechanical agitation and dilution in the generation of water memory? How can water and EM radiation represent genetic code?

A theoretical explanation of how the memory of water might work must still be explored. However, the fact that the effective transmission of molecular signals has now been observed by independent teams using different biological systems provides a strong additional basis to suggest that the phenomena observed by Benveniste are indeed valid. Whatever results ongoing and future investigations may bring, the difficult road Jacques Benveniste traveled, by opposing the automatic acceptance of received ideas, will have contributed to sustaining freedom in scientific research and placing emphasis, where it belongs: an observable fact.

Endnotes

Del Giudice E., Preparata G., Vitiello. G. *Phys Rev Lett* 1988, 61: 1085–1088

Thomas Y., Schiff, M., Belkadi, L., Jurgens, P., Kahhak, L., Benveniste, *J. Medical Hypotheses* 2000, 54: 33-39

Thomas, Y., Kahhak, L., Aissa, J. In, *Water and the Cell*, 2006, pp 325–340. Ed. GH Pollack, IL Cameron and DN Wheatley (Springer, Dordrecht)

Thomas, Y. *Homeopathy.* 2007, 96: 151–157

Albrecht-Buehler, G. *Proc Natl Acad Sci U.S.A.* 1992, 89:8288–8292; 2005, 102, 14: 5050–5055

Trushin, M.W. *Microbiology.* 2003; 149: 363–368

Ninham, B.W., Boström, M. *Cell Mol Biol.* 2005, 51 (8): 803–813

Ben Jacob, E., Aharonov, Y., Shapira, Y. *Biofilms* 2004, 239–263

Vallée Ph, Lafait J, Mentré P, Monod MO, Thomas Y. *J Chem Phys* 2005, 122: 114513–114521; Langmuir, 2005, 21(6): 2293–2299

Fels, D., *PLoS ONE*, 2009, 4(7), doi: 10.1371

Cifra, M., Fields, J.Z., Farhadi, A. *Progress in Biophysics and Molecular Biology* 2011, Vol. 105, Issue 3 :223–246

Chaban, V.V., Cho, T., Reid, C.B., Norris, K.C., *Am J Transl Res.* 2013, 5(1):69–79

Dunne, B.J., Jahn, R.G. *Cell Mol Biol* 2005, Dec 14; 51(7):703–714

Jonas, W. B., Ives, J.A., Rollwagen, F., Denman, D.W., Hintz, K., Hammer, M., Crawford C., Henry, K. *FASEB J.* 2006, 20, 23–28

Endler, P.C., Pongratz, W., Smith, C.W., Schulte, J. *Vet. Human Toxicol.* 1995, 37: 259–263

Endler, P.C., Heckmann, C., Lauppert, E., Pongratz, W., Alex, J., Dieterle, D., Lukitsch, C., Vinattieri, C., Smith, C.W., Senekowitsch, F., Moeller, H., Schulte, J. In, *Fundamental Research in Ultra High Dilution and Homoeopathy.* 1998, Ed. Schulte J, Endler PC (Kluwer, Dordrecht)

Montagnier, L., Aïssa, J., Ferris, S., Montagnier, J.L., Lavallée, C. "Interdiscip Sci Comput Life" Sci, 2009, 1: 81–90

Montagnier, L., Aïssa, J., Lavallée, C., Mbamy, M., Varon, J., Chenal, H. *Interdiscip Sci Comput Life Sci*, 2009, 1: 245–253

Montagnier, L., Aissa, J., Del Giudice, E., Lavallee, C., Tedeschi, A., Vitiello, G., *Journal of Physics*, 2011, J. Phys.: Conf. Ser. Vol. 306: 012007

Gariaev, P.P., Grigor'ev, K.V., Vasil'ev, A.A., Poponin, V.P., Shcheglov, V.A. *Bulletin of the Lebedev Physics Institute*, 1992 n. 11-12, p. 23–30

Gariaev, P.P., Marcer, P.J., Leonova-Gariaeva, K.A, Kaempf, U., Artjukh, V.D. *DNA Decipher Journal*, January 2011, Vol. 1, Issue 1, pp. 025–046

Butters, J.T., Figueroa, X.A., Butters, B.M. *Open Journal of Biophysics*, 2014; 4: 147–168.

The Unifying Property of Sound

THOMAS ORR ANDERSON

Abstract

In this chapter we explore the inherently unifying property of sound as it applies to a conscious being with a sound-sensitive nervous system and object recognition faculties. This unifying property underlies an essentially holistic framework for yielding effects upon a whole person, and thereby facilitates holistic approaches to health and wellness. These considerations suggest a variety of research directions toward novel applications.

Part 1: Introduction and Background

The great successes of modern science have been primarily the results of a reductionist approach by which we model the universe through acts of progressively dividing, dissecting, and compartmentalizing. As we divide our conceptions of the universe into ever smaller particles, forces, and mechanisms, we likewise divide our fields of study into ever more discrete areas of specialization. And as medical practices naturally follow the progressions of science, we also divide our conceptions of the human being into a complex array of disparate mechanisms and processes. Thus, we require one doctor for bones, another doctor for feet, another for depression, and so on.

Although the reductionist approach to health has undeniably yielded vast benefits, drastically extending lifespans and eliminating many once-pervasive illnesses, it also has demonstrated some fundamental shortcomings, but this approach is not well suited for treating the person as whole. In many cases, a holistic approach that simultaneously addresses the mental, emotional, and spiritual realms of experience can be more appropriate. Notable examples of such cases include psychosomatic illnesses, chronic pain, stress, depression, and general health and vitality.[1, 2]

Fortunately, in our search for holistic approaches, we may turn to traditional healing practices in which we find a plethora of time-tested principles and tools. One ubiquitous feature of traditional healing systems around the world is the use of sound. From the shaman's drums to the priest's prayers to the mantras of the mystic, sound has historically played a central role in the healing arts.[3]

Here we explore a simple reason for sound being uniquely useful in holistic healing approaches. Specifically, we examine the inherently unifying property of sound and how this property interfaces with the whole person. A clear understanding of this unifying property illuminates how we may knowledgeably utilize sound for our health and wellness, and thereby regain some unique benefits of traditional holistic approaches while losing none of the benefits offered by modern medical science.

Part 2: Sound and Unity

2.1: Sound Unifies Our Experience

Typically, we think of sound only as that which our ears sense, what we hear. However, our experience of sound is far more vast and encompassing. In fact, most of the human nervous system is sensitive to sound. The great majority of our 43 total pairs of cranial and spinal nerves by which we sense the world around us, include sensitivity to pressure variations. For this reason, many of the sounds we hear are simultaneously felt in our body. Likewise, many of the vibrations we feel are simultaneously heard with our ears. Sound thus unifies our perceptions and provides an opportunity to perceive one unified phenomenon by means of an extended array of sensory input channels.

Sound is also unifying in another important sense: affecting simultaneously our inner and outer worlds. We witness sound as an objective physical phenomenon in the form of vibrations affecting both our body and environment. But we also experience sound as an internal subjective phenomenon, affecting our emotions, conjuring memories, and interpenetrating the layers of our mind. Sound thus naturally unifies our consciousness in the sense that it bridges that which is internal or subjective and that which is external or objective.

2.2: Unity Inherent in Waves

Sound is not only unifying within our consciousness and nervous system, but by its very nature, its wave nature in particular. Let us examine a simple sine wave (Figure 1).

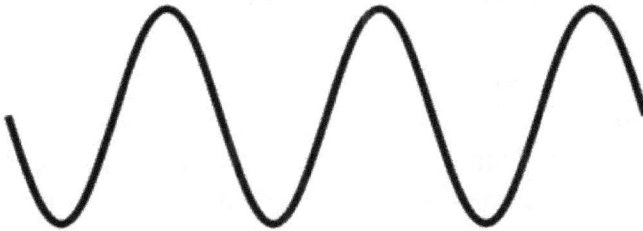

Figure 1 – A Sine Wave

Although the figure only shows three full cycles of the wave, we can easily imagine that it extends indefinitely far in either direction. Notice the symmetrical repetition inherent in this waveform. In fact, symmetry and repetition are characteristics of all waves including, of course, sound waves.

What are sound waves?

In a basic physical sense, sound waves are repetitive patterns of pressure or elastic stress variations in a physical medium. These variations travel as elastic motions in every direction, interacting with both the physical materials as well as with other sound waves. As sound waves cross one another within the medium, interference patterns emerge which possess extended unified structures. We shall avoid going into the detailed physics of sound, as there is already a plethora of excellent presentations on this extensive topic.[4,5] Rather, we shall only briefly examine the general structure of sound waves so as to expose their inherently unifying property.

2.3: 1-Dimensional Sound Waves

One of the simplest types of sound waves to visualize is a single frequency sound in a cylindrical cavity. This case can be treated as essentially one-dimensional for our purposes.

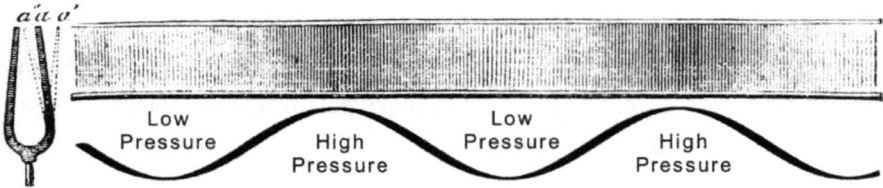

Figure 2 – Single Frequency Sound Wave in a Cylindrical Cavity

Figure 2 shows the repetitive symmetrical structure formed by a single frequency sound wave in a gas-filled cylinder, which gas can, of course, be air. As the tuning fork moves back and forth, it pushes and pulls the gas molecules rhythmically. This rhythmic motion causes a stable unified pattern of high- and low-pressure regions to emerge. Plotting the pressure variations along the length of the cylinder yields a simple sine wave.

What we would like to emphasize here is the unifying action of this process. Prior to introducing a sound wave, the gas molecules in the cylinder essentially move randomly, each molecular motion practically independent of all others. Once a sound is introduced, the molecular motions collectively form a stable extended symmetrical structure in which the motions of distant molecules are now distinctly interdependent. Each individual molecule is now vibrating at the same frequency as all the others. Thus, the entire cylinder of gas is unified by the sound wave into a coherent whole.

2.4: 2-Dimensional Sound Waves

The symmetrically unifying action of sound upon physical structures is vividly illustrated in Chladni patterns formed in powders on thin vibrating plates. These patterns allow us to easily visualize the more complex two-dimensional sonic structures.

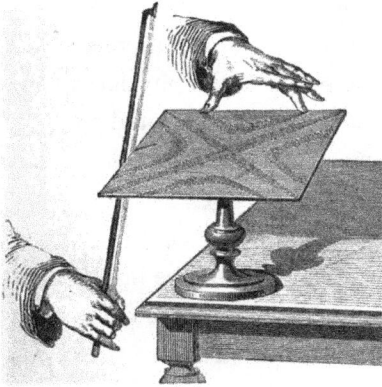

Figure 3 – Creating Chladni Patterns

Figure 3 shows a violin bow rubbing against a thin metal plate. This plate will generally be covered in sand or some other fine powder that can move easily across the surface. If the bowing vibrates the plate at one of its resonant frequencies (modes), a symmetrical interference pattern of nodes and anti-nodes emerges. The nodes are those regions where the plate moves the least (maximum *destructive interference*). The anti-nodes are those regions where the plate moves most (maximum *constructive interference*). The powder is naturally scattered away by the motion at the anti-nodes and collects in the stationary nodes. In this way, the structure of the resonant sound wave is made beautifully visible in the powder's subsequent arrangement on the plate.

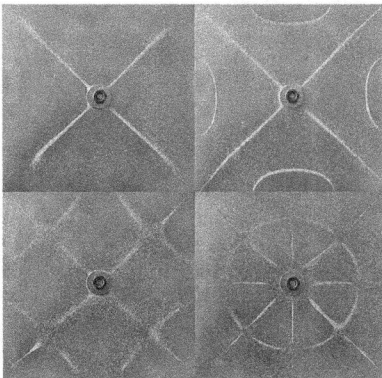

Figure 4 – Chladni Patterns

NOTE: It is important to mention here that every vibrating physical object has a practically infinite number of resonant frequencies (modes). The structures formed by these modes grow in complexity with higher frequencies as shorter wavelengths permit finer resolution.

Much like the gas molecules in the cylinder of our previous example, the particles of powder may be randomly placed on the plate prior to the introduction of sound, such that each particle's location is initially independent of all others. As sound is introduced to the system, a holistic unity emerges wherein the positions and motions of all the particles become symmetrically interdependent. Here again, we see the inherently unifying property of sound waves at work.

2.5: 3-Dimensional Sound Waves

Three-dimensional sonic resonance patterns produce even more visually striking images. So inspired by these images was Swiss physician Hans Jenny, that in 1967 he proposed a new field of science devoted solely to their study.[6] Based upon the Greek word κῦμα (pronounced *kýma*), meaning "wave," he named this new field *Cymatics*.

Figure 5 is a typical Cymatics photo of an illuminated circular container of water vibrating at one of its resonant modes. As the light reflects from the water's undulating surface, the resonance pattern is made beautifully visible, revealing an elaborate symmetrical structure within the water.

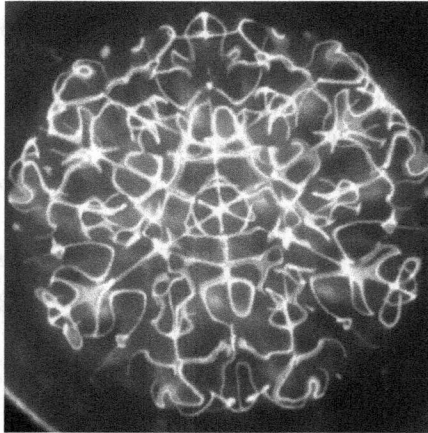

Figure 5 – Cymatics Image of Sound in a Circular Container of Water

Just as in the previous examples, the water's molecular motions are relatively independent prior to the introduction of sound, moving essentially randomly in every direction. But with the introduction of sound, the molecular motions become highly interdependent. Since our bodies are composed primarily of water, this particular example holds special significance for us. The principle is general, however, and applies to the action of sound in any substance.

Part 3: Sound and the Human Nervous System

In Part 2, we explored the inherently unifying property of sound. In order to better understand how this property applies to the human experience, we must also consider how sound interacts with our nervous system.

3.1: Sound and the Human Nervous System

For our purposes, it is not necessary to go into great detail regarding nervous system structures and processes. Rather, we require only a simple illustration of the human nervous system as it employs a sensor array that transmits information about environmental conditions to an object recognition system (*ORS*).

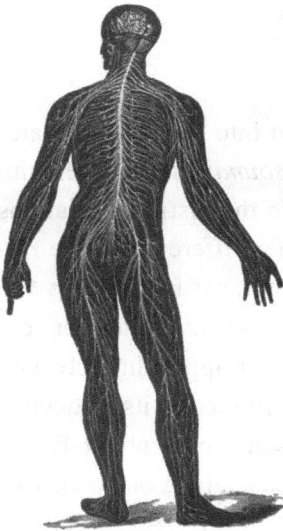

Figure 6 – The Human Nervous System

Figure 6 roughly illustrates the human nervous system, including the brain, spinal cord and the 43 total pairs of nerves that branch dendritically through the body. It is worth noting that although the human body is immensely complex, all sensory information reaches the brain via only this small number of nerve channels. The nerve channels transmit information bi-directionally between the brain and the body/environment. For our purposes, however, we need only consider one direction, wherein signals carry environmental sensory information to our brain for processing.

Each of the 43 nerve pairs emanating from the spinal cord and cranium receives signals from a limited region of the body. Some of the nerves branch internally amongst the organs, muscles, and other subsystems, and some to the body's outer surface areas, near and on the skin. There are specialized receptors at the end of the nerve branches that interact with the environment, including the body itself, and transmit sensory information into the nerve channels via complex electro-chemical processes. This information is then conducted by the nerve channels to the brain.

The human nervous system employs a variety of receptors, sensing a broad range of conditions that include temperature, pain, pleasure, etc. In our case, we are only interested in those receptors that are sensitive to sound. As it turns out, the vast majority of the 43 nerve pairs carry information about our sonic environment.[7]

The 43 nerve pairs are divided into two distinct groups. From the spine emanate the 31 pairs of *spinal nerves*, each of which receives signals from a variety of sound-sensitive receptors.[8] The skull contains the other 12 pairs of nerves. These *cranial nerves* branch out primarily around the head, neck and shoulders. At least 3 of these cranial nerve pairs also lead to sound-sensitive receptors, including, of course, the ears.[9] Thus, the human body is largely a hearing system, sensing both tactile vibrations all over the body and air pressure variations at the ears.

3.2: The Two Kinds of Hearing

The scope of human hearing is generally divided into two distinct categories: sounds we hear with our ears and *tactile sounds* that we feel with our body. This division is natural not only due to the distinct variations in how we experience the sensations, but also to a difference in the frequency responses of the two categories. The human ear is responsive to vibrations in the range of approximately 20 to 20,000 Hz (cycles per second). The body, however, is sensitive to a range of approximately 1 to 3000 Hz. Although the tactile bandwidth is more limited in its frequency range—it covers at least 11 octaves—the ears sense only about 10 octaves. (In contrast, the human eye is only sensitive to about one octave of the electromagnetic spectrum.)

Part 4: Mr. Hand — An Artificial Conscious Being

4.1: Developing a Model

In order to better understand how the unifying property of sound interacts with a person, and particularly how sound may be useful in holistic healing strategies, we introduce a simple model of a conscious being, whom we playfully name Mr. Hand. We develop this model in rudimentary detail, including only those features necessary for our considerations.

Note: We refrain from utilizing any specific definitions or theories of consciousness so as to maintain the generality of our considerations.

For our purposes, we only require that *Mr. Hand*

- is sentient
- lives in a space-time universe
- perceives his surroundings via a sound-sensitive sensor array
- employs object recognition processing of sensory information in the act of perception

4.2: The Sensor Array

Let us first develop a model of *Mr. Hand's* sensor array as a conveniently simplified version of the human hand (Figure 7).

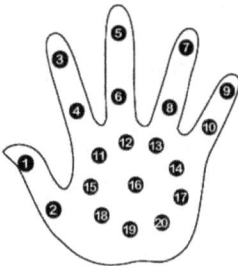

Figure 7 – Model Hand—a 20-sensor Array

Our model hand has a total of 20 pressure-sensitive artificial nerves endings that we shall call sensors. For simplicity, we limit their sensitivity to only one environmental condition: pressure (P). The signal measured by each sensor can be denoted as a function of time:

$$P(t)_n \equiv pressure\ at\ the\ n_{th}\ sensor\ at\ time\ t \quad (Eq.1)$$

Figure 8 – Single Sensor Contact

Let us also assume that the sensors are capable of transmitting their pressure measurements with no noise, signal loss, or resolution limitations. The precise nature of the pressure measurement mechanisms and the signal-encoding transmission processes is not required for our model.

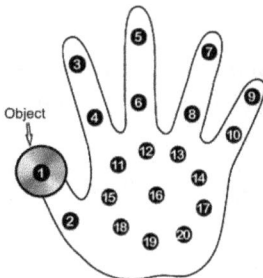

4.3: Single Sensor Contact

Suppose that only the tip of the thumb makes temporary contact with an object (Figure 8).

If the contact is made at time $t = 1$ and lasts for one second, the measurement made by S_1 can be graphically represented as:

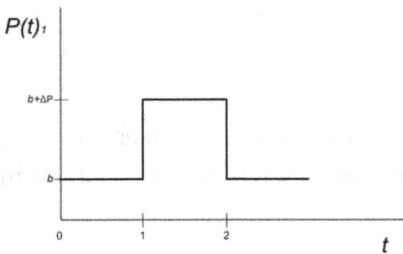

Figure 9 – Sensor 1 Signal

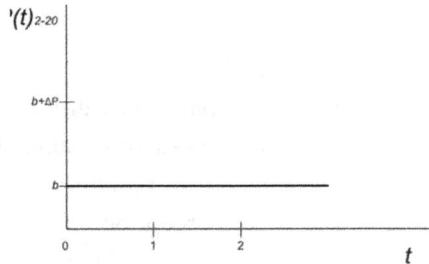

Figure 10 – Sensors 2-20 Signal

where b is the baseline pressure reading in the absence of object contact and ΔP is the magnitude of the applied pressure.

The measurements made by all other sensors (S_{2-20}) in this case can be represented as the difference between the pressure measurements at S1 and the measurements at all other sensors, as made clearly visible in the above graphs, can be readily utilized by Mr. Hand's object recognition faculties to detect the object. For our considerations, we need not specify the precise means by which the signals are compared in the object recognition process, as this can be achieved with a variety of simple methods.[10]

4.4: Extensive Object Contact

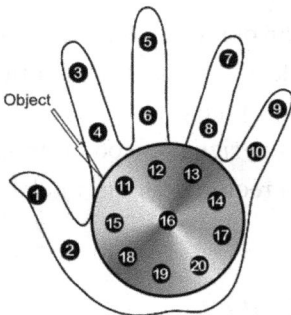

Figure 11 – Extensive Object Contact

Let us now suppose that the hand makes contact with a more extensive flat object such that a pressure of equal magnitude (ΔP) is simultaneously applied to an extensive subset of the 20-sensor array:

We shall assume that the pressure is instantaneously applied and released and is present at times $t = 1$ and $t = 2$, but not before $t = 1$ and not after $t = 2$.

The pressure at the sensors is in this case given by:

$$P(1 \leq t \leq 2)_{n\,=\,11\text{-}20} = b + \Delta P \quad (Eq.\,2)$$

$$P(t < 1)_{n\,=\,11\text{-}20} = P(t > 2)_{n\,=\,11\text{-}20} = P(t)_{n\,=\,1\text{-}10} = b \quad (Eq.\,3)$$

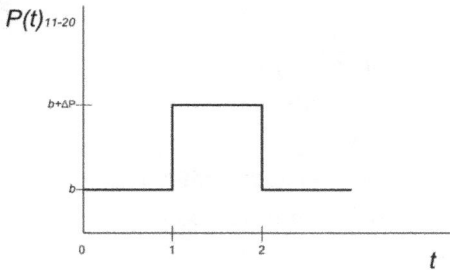

Figure 12 – Sensors 11-20 Signal

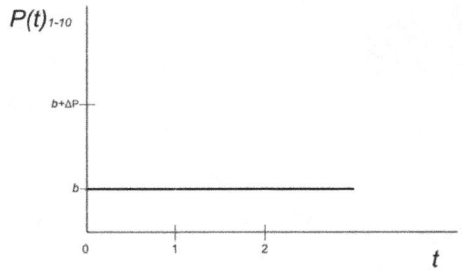

Figure 13 – Sensors 1-10 Signal

These signal variations may also be utilized by *Mr. Hand's* object recognition faculties to detect this more extensive object.

4.5: Object Recognition

So as to maintain the general applicability of our considerations, we assume that *Mr. Hand* employs a simple *object recognition system (ORS)* that performs only one primary action: it detects coherent patterns in the signals from the sensor array and labels these patterns as objects. Thus, a sort of internal perceptual universe is created within *Mr. Hand* which we shall denote as U_H. U_H is essentially Mr. Hand's internal model of the universe, composed of those patterns labeled as objects, in a background of *not-objects*, which we denote as *NO*.

As this internal world of perceived objects can be contrasted with the external world that is being sensed, we can refer to the *subjective* and *objective* experiences of *Mr. Hand*. We shall denote the physical description of the object contact as his objective experience. The internal experience of *Mr. Hand's* consciousness, as he employs his *ORS*, shall be denoted as his *subjective* experience. Let us now examine the two previous examples in this context.

In the first case, where a small object only contacts one sensor, *Mr. Hand's* objective and subjective experiences may be summarized as:

Objective experience (small object)
A pressure of magnitude ΔP was applied to S_1 from time t = 1 until time t = 2, at which point the applied pressure ceased.

Subjective experience (small object)
My thumb touched a small object for about a second.

In the second case, where the hand makes contact with a more extensive flat object, *Mr. Hand's* experience may be summarized as:

Objective experience (extensive object)
A pressure of magnitude ΔP was applied to Sensors 11-20 from time t = 1 until time t = 2, at which point the applied pressure ceased.

Subjective experience (extensive object)
My entire palm touched an object for about a second.

4.6: Mr. Hand's Universe
For our considerations, we need not know the exact mechanisms and algorithms involved in *Mr. Hand's ORS*. Rather, we require only a simple topological examination of some basic relationships between three distinct sets:

1. The Total Universe (U_T)
2. *Mr. Hand's* Perceived Universe (U_H)
3. The Object (O)

Note: We are using O to designate both the labeled object perception as well as the physical object being perceived. This simplification is adequate for our purposes.

As *Mr. Hand's* sensors are only receptive to pressure, his perceived universe (U_H) contains only one type of object: *pressure-objects* embedded in space and time (static *pressure-objects* are the limiting case of sound-objects with frequency = 0). If, as we shall reasonably assume, the total universe (U_T) includes phenomena other than pressure variations, *Mr. Hand's* perceived universe (U_H) is at most only a subset of the total universe (U_T) (Figure 14).

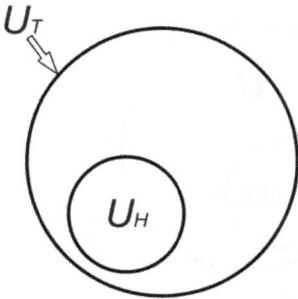

Figure 14 – Perceived Universe as a Subset of the Total Universe

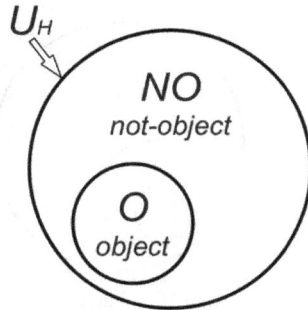

Figure 15 – Object Recognition as a Binary Division of the Perceived Universe

Just as U_H is a subset of the total universe (U_T), the detected object (O) is itself a subset of U_H. Thus, as noted above in section 4.5, U_H is divided into the two distinct subsets: object (O) and not-object (NO) (Figure 15).

4.7: Pervasive Object Contact

Let us now consider a sequence where *Mr. Hand's* sensor array contacts progressively larger objects. In particular, we consider a discrete progression of 20 object sizes, such that *Object 1 (O_1)* is only extensive enough to contact one sensor, *Object 2 (O_2)* is only extensive enough to contact two sensors and so forth such that *Object n (O_n)* is only extensive enough to contact *n* sensors.

Suppose that the sensor array makes temporary contact with each of these objects, progressively from O_1 to O_{20}. *Mr. Hand* would, in this case, detect progressively larger objects until the largest, O_{20}, which would completely fill his perceived universe (U_H) (Figure 16).

In the first 19 cases, upon detecting the object, *Mr. Hand's ORS* divides his perceived universe (U_H) into the distinct subsets, On and NOn. But in the case of the largest object O_{20}, U_H is not divided. Rather, in this limiting case, U_H is entirely filled with one single object, O_{20}. Therefore, detecting such a pervasive object with all sensors simultaneously, essentially unifies *Mr. Hand's* perceived universe (U_H), eliminating the possibility of the object / not-object distinction. As we will show, this case is particularly relevant when *Mr. Hand* is immersed in sound waves.

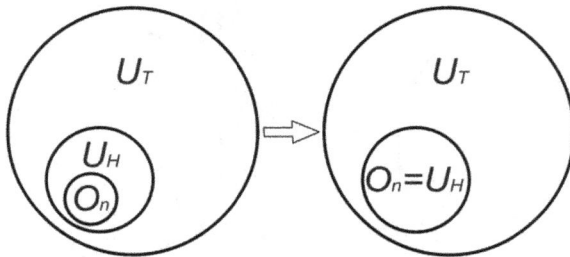

Figure 16 – Detection of Pervasive Object Eliminates Binary Division of Perceived Universe

4.8: Vibrating Object Contact

Thus far we have explored only *Mr. Hand's* detection of static flat objects that temporarily apply equal pressure to some subset of his 20 sensors. The detection of these objects only requires that his *ORS* be capable of sensing pressure amplitude variations. Let us now consider the case of vibrating objects. In order for *Mr. Hand's ORS* to detect such objects, it is reasonable to employ some form of spectral analysis. With the addition of spectral analysis capabilities, his *ORS* can detect not only patterns in pressure magnitudes at the sensors but can also detect patterns in the frequency composition of sensor fluctuations over time. This addition facilitates detection of objects that do not necessarily contact sensors simultaneously and thus provides additional sensitivity to objects with vibrating surfaces of contact. We shall assume that *Mr. Hand's ORS* indeed does employ spectral analysis and see how this applies to our considerations.

Suppose *Mr. Hand's* sensor array contacts a vibrating object at one sensor. In this case, just as in the case of static object contact, his *ORS* divides U_H into the subsets O *and* NO. Likewise, when the array detects a more extensive vibrating object, making contact with multiple sensors, his *ORS* also divides U_H into those same distinct subsets.

4.9: Sound Wave Contact

In the previous section, we equipped *Mr. Hand's ORS* with spectral analysis capabilities, such that he can now detect vibrating objects as well as static objects. This addition also affords him another new ability that is particularly relevant to our considerations: *Mr. Hand* is now capable of detecting fields of mechanical vibrations — sound waves. The object is no longer required to have a well-defined boundary. It can be simply a sound wave reaching his sensor array. If we assume that the sound wave is pervasive, such that it contacts all of his sensors, *Mr. Hand's* entire perceived universe (U_H) will be united as one coherent whole (Figure 17). Here we see a most pertinent example of the inherently unifying property of sound: *Sound immersion unifies Mr. Hand's entire perceived universe as one coherent structure of nonlocal relationships.*

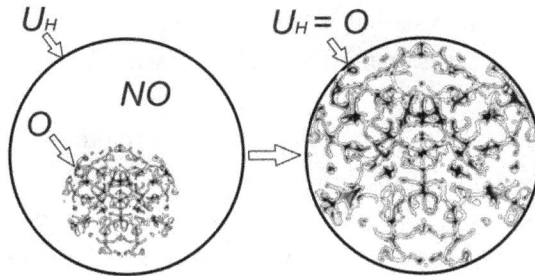

Figure 17 – Sound Immersion Eliminates Binary Division of Perceived Universe

In light of this important result, let us conclude by returning our considerations to the human experience, and leave *Mr. Hand* behind.

Part 5: Summary and Conclusion

It will be useful to summarize what we have covered in these explorations. Here are the primary points of our considerations:

- The reductionist approach of modern scientific and medical practices has not only yielded vast benefits but has also unfortunately overshadowed some benefits unique to holistic approaches.
- Sound has a fundamentally unifying affect upon the whole-person.
  - Immersive sound unifies our perceptions as it is detected simultaneously via the vast majority of our nervous system's 43 pairs of input channels (cranial and spinal nerves).
  - Sound unifies our subjective and objective worlds, affecting simultaneously both our internal experiences (mental, emotional, etc.) and our physical environment (the physical structures of our bodies and surroundings).
- The unifying nature of sound is made beautifully visible as it forms symmetrical patterns in 1, 2, and 3-dimensional media.
- Humans can feel as well as hear sounds.
- Sound immersion can lead a conscious being, equipped with spectral analysis-based object recognition faculties, to perceive the universe as one unified coherent whole.

What may we learn from this somewhat circuitous exploration?

The author suggests the following:

Sound is a uniquely useful medium for holistic approaches to health and wellness as it naturally unifies our entire being... our consciousness and our biology.

Part 6: Research and Applications

The natural environments in which humans evolved primarily provided experiences of holistic interdependent phenomena, as interdependence is the very nature of ecological systems.

Unfortunately, the conveniences of modern technological innovations largely provide experiences of discrete independent phenomena, which experiences tend to disintegrate our mind-body connection. This synthetic mind-body disintegration can manifest negatively in a broad variety of contexts, notably in the case of our mental and physical health.

As we have demonstrated, sound is a uniquely effective medium for reestablishing and strengthening the mind-body connection. Notably, immersive sound experiences, wherein a coherent sound field acts upon the entire body and ears simultaneously, provide a uniquely holistic venue in which to yield effects upon a whole person. For this reason, immersive sound-therapy techniques may find strong support, offering inherent whole-person unifying capabilities.

Unfortunately, sound-therapy research has thus far remained generally quite limited. This situation is beginning to change, however, as ever more people are demonstrating positive health results from holistic approaches seldom achieved through reductionist methods. Here we suggest a few avenues of sound-therapy research that may facilitate this welcome change:

- The effects of sound immersion upon
  - the mutual coherence of bio-rhythms (brain waves, breathing, HRV, etc.)
  - acute and chronic pain
  - co-ordination
  - spatial awareness
  - chronic stress
  - wound healing
  - dissociative disorders
  - cancer reduction/growth
  - DNA replication and telomere length
  - lucid dreaming
  - OBE (out-of-body experience) occurrence
- The relationships between tactile sound and bio-energy (Qi, Prana, etc.)
- The effects of various frequencies and frequency combinations in sound immersion

Image Sources
Fig. 2: Popular Science Monthly Volume 13, 1878, via Wikimedia Commons.
Fig. 3: William Henry Stone (1879) Elementary Lessons on Sound, Macmillan and Co., London, p. 26, fig. 12, via Wikimedia Commons.
Fig. 4: Elmar Bergeler, www.science-niblets.org, via Wikimedia Commons.
Fig. 5: Hladnison, 2015, via Wikimedia Commons.
Fig. 6: Alvin Davison (1908) The Human Body and Health, American Book Company, via Wikimedia Commons.

Endnotes
1 "Mind-Body Medicine: State of the Science, Implications for Practice." *Journal of the American Board of Family Medicine*, March 1, 2003 vol. 16 no. 2 131–147.
2 Goleman, Daniel. *Mind Body Medicine*, Consumer Reports Publishing, 1998.
Burnett, Charles. "'Spiritual Medicine': Music and Healing in Islam and its Influence in Western Medicine." In *Musical Healing in Cultural Contexts*, Penelope Gouk, ed. 85–91. Aldershot and Brookfield: Ashgate, 2000.
Cook, Pat Moffitt. "Sacred Music Therapy in North India." *The World of Music* 39, no.1 (1997): 61–84.
Friedson, Steven. *Dancing Prophets: Musical Experience in Tumbuka Healing.* Chicago and London: The University of Chicago Press, 1996.
Gouk, Penelope, ed. *Musical Healing in Cultural Contexts.* Aldershot and Brookfield: Ashgate, 2000.
Horden, Peregrine, ed. *Music as Medicine: The History of Music Therapy Since Antiquity.* Aldershot: Ashgate, 2000.
Janzen, John M. "Theories of Music in African Ngoma Healing." In *Musical Healing in Cultural Contexts.* Penelope Gouk, ed. 45–66. Aldershot and Brookfield: Ashgate, 2000.
Koen, Benjamin D. "Musical Healing in Eastern Tajikistan: Transforming Stress and Depression Through Falak Performance." *Asian Music.* 37, no.2 (2006): 58–83.
Koen, Benjamin D., ed. *The Oxford Handbook of Medical Ethnomusicology.* New York: Oxford University Press, in press.
Larco, Laura. "Encounters with the Huacas: Ritual Dialogue, Musical Healing in Northern Peru." *The World of Music* 39, no.1 (1997): 35–60.
Meinecke, Bruno. "Music and Medicine in Classical Antiquity." In *Music and Medicine.* Schullian and Schoen, ed. 47-95. Freeport: Books for Libraries Press, 1948.
Petran, Laurence A. "Anthropology, Folk Music and Music Therapy," *Music Therapy* (1953): 247–50.
Radin, Paul. "Music and Medicine among Primitive Peoples." In *Music and Medicine.* Schullian and Schoen, ed. 3–24. Freeport: Books for Libraries Press, 1948.
Rochbacher, Michael J. "The Ethnomusicology of Music Therapy." Ph.D. diss., University of Maryland Baltimore County, 1993.
Roseman, Marina. *Healing Sounds from the Malaysian Rainforest: Temiar Music and Medicine.* Berkeley and Los Angeles: University of California Press, 1991.
4 Stork, David et al. *The Physics of Sound*, Pearson Publishing, 2004.
5 Benade, Arthur. *Fundamentals of Musical Acoustics*, Dover Publications, 1990
6 Jenny, Hans. *Kymatik Cymatics: The Structure and Dynamics of Waves and Vibrations*, Basilius Presse, 1967.

7 Kiernan, John et al. *The Human Nervous System*, LWW Publishing, 2013.

    Mai, Juergen et al. *The Human Nervous System*, Academic Press, 2011.

    Nieuwenhuys, Rudolf et al. *The Human Central Nervous System*, Steinkopff Publishing, 2007.

8 Website: https://en.wikipedia.org/wiki/Cranial_nerves

9 Website: https://en.wikipedia.org/wiki/Spinal_nerve

10 Bishop, Christopher. *Pattern Recognition and Machine Learning*, Information Science and
    Statistics, 2007.

    Grauman, Kristen et al. *Visual Object Recognition*, Morgan and Claypool Publishers, 2011.

    Tarr, Michael et al. *Object Recognition in Man*, Monkey, and Machine, MIT Press, 1999.

    Treiber, Marco Alexander. *An Introduction to Object Recognition*, Springer Publishing, 2010.

Intuitive Intelligence

Rollin McCraty

Abstract

This chapter explores the heart's role in emotional experience and one's intuitive capacities and access to a source of deeper wisdom. The implications of accessing intuition in meeting the increasingly complex demands of life with greater love, compassion and kindness, thereby raising consciousness, is discussed as well as the nature and types of intuition.

Introduction

Although not yet well understood, intuition is a familiar facet of experience that can be a powerful transforming mediator and can inform and potentially redirect the path of an individual's life. Intuitive perception is commonly acknowledged to play an important role in business decisions and entrepreneurship, learning, creativity, medical diagnosis, healing, social cognition, spiritual growth and overall well-being.[1,2] Yet despite its presence in informing such diverse aspects of individual and collective life, intuition is not well understood and largely remains a scientific mystery. Along with other scientists who deem these phenomena worthy of rigorous investigation, we believe that intuitive perception involves the heart, brain and nervous system's connection to a field of information beyond normal conscious awareness.[3-10] Our previous studies suggest that no matter how intuitive information is initially introduced into the psychophysiological systems, once received, it is processed in the same way that information obtained through the familiar sensory systems is processed.[10] As discussed in more detail in this chapter, we propose there are three types or categories of intuition: implicit knowledge, energetic sensitivity, and nonlocal.[11] We also suggest that an important aspect of self-regulatory capacity is aligning with one's innate higher-order wisdom or intuition and that without this alignment people are at risk of living their lives primarily through the filters of past experience.

Self-Regulation

Much attention has been given to identifying the many factors that go into making good decisions. Among these factors are awareness of self and others, cognitive flexibility, and self-regulation of emotions. Another factor that should be considered in decision-making and self-regulation, one we've all experienced, perhaps without being fully aware of it, is *intuition.*

Adjusting or self-regulating one's responses and behavior in order to build and maintain loving relationships and a supportive social network and effectively meet life's demands with composure, consistency and integrity, arguably is central to good health and effective decision-making.[12] Despite the importance of self-directed control, many people's ability to self-regulate is far less than ideal. In fact, failures of self-regulation, especially of emotions and attitudes, are central to the vast majority of personal and social problems that plague modern societies. Therefore, the most important skill the majority of people need to learn is how to increase their capacity to self-regulate emotions, attitudes and behaviors. Self-regulation enables people to mature and meet the challenges and stresses of everyday life with resilience. We suggest that an under-recognized aspect of self-regulation is aligning with one's innate higher-order wisdom or intuition. Without this alignment with our intuitive guidance, people are at risk of living their lives primarily through the automatic filters of past familiar experience.

Intuition

The root of the term 'intuition' stem from the Latin word *in-tuir*, which can be translated as 'looking, regarding or knowing from within.' In his article reviewing intuition, Gerard Hodgkinson concludes that "intuiting" is a complex set of interrelated cognitive, affective and somatic processes, in which there is no apparent intrusion of deliberate, rational thought. He also concludes that the considerable body of theory and research that has emerged in recent years clearly demonstrates that the concept of intuition has emerged as a legitimate subject of scientific inquiry that has important ramifications for educational, personal, medical and organizational decision making, personnel selection and assessment, team dynamics, training and organizational development.[13] Another comprehensive review of the intuition literature defined intuition as "affectively charged judgments that arise through rapid, non-conscious and holistic associations."[14] Neuroscientist Antonio Damasio also suggests that the outcomes of intuition can be experienced as a holistic "hunch" or "gut feel," a sense of calling or overpowering certainty and an awareness of a knowledge that is on the threshold of conscious perception.[15,16]

Several researchers have contended that intuition is an innate ability that all humans possess in one form or another and is arguably the most universal natural ability we possess. They also say the ability to intuit could be regarded as an inherited unlearned gift.[17,18] A common element also found in most discussions and definitions of intuition is that of affect or emotions. Although intuitions are felt, they can be accompanied by cognitive content, an inner voice and perception of previously unknown information. We suggest that emotions are the primary language of intuition and that a specific type of intuition offers a largely untapped resource to manage and uplift our emotions, our daily experience and our consciousness.

Types of Intuition

For purposes of this discussion, I will use the framework generated by our research at the HeartMath Institute, which suggests there are three different categories or types of processes that are often used to describe intuition, and will discuss the relevancy of each type.

Implicit Learning Processes

The first type of intuition is often called *implicit knowledge* or implicit learning. The majority of studies on intuition have been grounded in implicit processes and confine intuition to a function of the unconscious mind accessing existing but forgotten information stored in the brain.[1,19-23]

From this perspective, awareness is thought to be restricted to perceptions of present sensory input, intermingled with memories of the past. Research in the fields of cognitive and social psychology has produced the commonly accepted dual-process theory. Although dual-process theories come in a number of forms, they all have in common the distinction of including two separate processing systems. The first system is contextually dependent, associative, heuristic, tacit, intuitive and implicit/automatic in nature. It processes information very rapidly and associates current inputs to the brain with past experiences. In this regard, the brain matches the patterns of new problems or challenges with implicit memories based on prior experience.[13,24,25] It is therefore relatively undemanding in terms of its use of cognitive resources. For example, when individuals have gained experience in a particular field, implicit intuitions are derived from their capacity to recognize important environmental cues and rapidly and unconsciously match those cues to existing familiar patterns. This results in rapid diagnosis or problem-solving. In contrast, the second processing system is contextually independent, rule-based, analytic and explicit in nature. It is relatively slow and places greater demands on cognitive resources than the first system.[13]

As these types of intuition are based on the past associations, if the past reference was not the best or accurate solution or perception for the current situation they can often lead to ineffective perceptions.

The implicit processes are also involved in what is commonly referred to in the scientific literature as insight. When we have a problem we cannot immediately solve, the brain can be working on it subconsciously. It is common when we are in the shower, driving, or doing something else, for example, and we're not thinking about the problem, that a solution pops into the conscious mind that we experience as an intuitive insight. This type of implicit process involves a long gestation period following an impasse in problem solving prior to a sudden insightful perception or strategy that leads to a solution.[26] In contrast, intuition in the first process described above occurs almost instantaneously.[22]

Energetic Sensitivity

The second type of intuition is what we call *energetic sensitivity*, which refers to the ability of our nervous system to detect and respond to environmental signals such as electromagnetic and biofields.[27,28] It is well established that in both humans and animals, nervous-system activity is affected by geomagnetic activity.[29,30] Some people, for example, appear to have the capacity to feel or sense that an earthquake is about to occur before it happens. It has recently been shown that changes in the earth's magnetic field can be detected about an hour or even longer before a large earthquake occurs.[31] Another example of energetic sensitivity is the sense that someone is staring at us, and several scientific studies have verified this type of sensitivity.[32]

To further understand energetic sensitivity, consider that the interaction and communication between human beings such as a mother and young child, or the consultation between patient and clinician, is a very sophisticated dance that involves many subtle factors. In addition to words, most people tend to think of communication solely in terms of overt signals expressed through facial movements, voice qualities, gestures and body movements. However, evidence now supports the perspective that a subtle yet influential magnetic or "energetic" communication system operates just below our conscious level of awareness that we can experience as a type of energetic sensitivity that is likely an aspect of empathy.[33]

The ability to sense other people's inner states is an important factor in allowing us to connect or communicate effectively with others. The ease or flow in social interactions depends to a great extent on establishing a spontaneous entrainment, or connection between individuals. When people are engaged in deep conversation, they can begin to fall into a subtle energetic dance, synchronizing their movements and postures, vocal pitch, speaking rates, and length of pauses between responses as they subconsciously sense other's inner states.[34] Additionally, as we are now discovering, important aspects of their physiology also can become linked.[27] One such example of energetic communication was provided from an experiment we conducted to investigate the energetic exchange and physiological linkage between people seated 5 feet apart. We used signal-averaging techniques to detect signals that were synchronous with the peak of the R-wave of one participant's electro-cardiogram (ECG) and recordings of another participant's electroen-cephalogram (EEG), or brain waves.

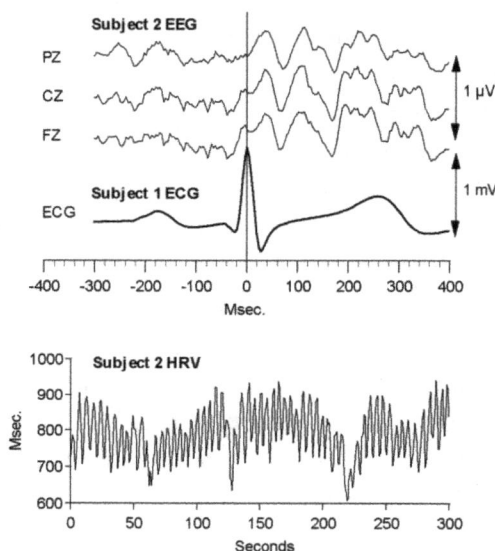

Figure 1 – Heart-brain synchronization between two people
The top three traces are Subject 2's signal-averaged EEG waveforms, which are synchronized to the R-wave of Subject 1's ECG. The lower plot shows Subject 2's heart rate variability pattern, which was coherent throughout the majority of the record.

Figure 1 illustrates an example of the synchronization of a study of a participant's brain waves to another participant's heartbeats (ECG signal). Power spectrum analysis of the signal-averaged EEG waveforms showed that the alpha rhythm was synchronized to others' heartbeats. In this example, when the data from the same two participants was analyzed to see if the other participant's brainwaves (EEG) were also synchronized to other person's heartbeats, there was no observable synchronization. The key difference between the participants was the high degree of cardiac coherence maintained by the participant whose brainwaves were synchronized to the other person's heartbeats. In other words, the degree of an individual's heart coherence (discussed later) appears to be a key factor in linking to others' physiological activity and rhythms, which essentially means having information about another person's current emotional states. This suggests that when one is in a heart-coherent mode, it facilitates a greater level of energetic sensitivity and potential for empathic awareness.

A study of a Spanish fire-walking ritual also demonstrated physiological synchronization between people. Heart-rate data was obtained from 38 participants, and synchronized activity between fire-walkers and spectators was compared. They found finely tuned synchronizations during the fire-walking ceremony between the fire-walkers and spectators, who were family members or had close personal relationships to them, but not with unrelated spectators. The authors suggested that the mediating mechanism was likely informational in nature.[35]

Steve Morris studied the effect of heart coherence in a group setting, conducting 148 trials with groups of four participants seated around a table.[36] Three of the participants at each table were recently trained in shifting into and sustaining a coherent state, and the fourth was untrained and unaware of the study's purpose. Morris wanted to determine whether the trained participants could collectively facilitate higher levels of HRV coherence in the untrained individual. He found that the untrained participants' coherence was higher in approximately half of all matched comparisons when the trained participants were in coherent states. He also found evidence of complex heart-rhythm synchronizations between all of the group participants, and also that higher levels of this synchronization correlated with relational measures (bonding) among them. Additionally, he found evidence of heart-to-heart synchronization across subjects, lending credence to the possibility of "heart-to-heart biocommunications." This study also suggested that a "group field" is formed that connects the group members where intuitive information is shared among the members of the group.

Nonlocal Intuition

The third type of intuition is *nonlocal intuition*, which refers to the knowledge or sense of something that cannot be explained by past or forgotten knowledge or locally occurring environmental signals. Examples of nonlocal intuition include when a parent accurately senses something is happening to his or her child who might be many miles away; remote viewing;[37] or the repeated, successful sensing experienced by entrepreneurs about factors related to making effective business decisions.[38]

While there are various theories that attempt to explain how the process of intuition functions, these theories have yet to be confirmed, so an integrated theory remains to be formulated. Nevertheless, there is increasing evidence showing that nonlocal intuition is a very real and measurable phenomenon. It has been suggested that the capacity to receive and process information about nonlocal events appears to be a property of all biological organization and is likely due to the inherent interconnectedness of everything in the universe.[4,5,7] There is now a large body of rigorous experimental research, dating back more than seventy years, documenting nonlocal intuitive perception in rigorous scientific experiments showing that it *cannot* be explained by flaws in experimental design or research methods, statistical techniques, chance or selective reporting of results.[37,39] For example, Bem conducted a meta-analysis of nine experiments in the area of precognition (conscious cognitive awareness) and premonition (affective apprehension) of a future event that could not otherwise be anticipated through any known sensory process, and found significant results in eight of the nine studies with respect to pre-stimulus responses in over 1,000 subjects, indicating a possible retroactive influence of the stimulus.[40] Mossbridge also recently concluded a meta-analysis of 26 physiological-based studies, which found a clear physiological pre-stimulus effect of what appeared to be unpredictable stimuli, despite the fact there was no known explanation available for this finding.[41]

One important conclusion from these studies is that intuitive perception of a future event is related to the degree of emotional significance of that event.[42,43] Moreover, the response to and processing of pre-stimulus information about a future event is not confined to the brain alone. Instead, the evidence from studies in our laboratory suggests that the heart responds first and then the brain and possibly other organs in the body, and all are involved together in responding to nonlocal intuitive information.[10,43]

The first studies we are aware of that examined changes in brain activity that preceded an unknown stimulus were conducted by Levin and Kennedy in the mid-70s.[44] They observed a significantly larger contingent negative variation (CNV), which is a slow brain wave potential associated with anticipation, expectancy, or cortical priming, just before subjects were presented a target stimulus. Warren et al. also found significant differences in event-related potentials (ERP) between target and non-target stimuli presented during forced-choice precognition tasks.[45] Don et al. extended these ERP findings in a series of gambling studies in which they found enhanced negativity in the ERPs was widely distributed across the scalp in response to future targets.[46,47] The authors concluded from these studies that the ERP effect was an indicator of "unconscious precognition" since the study's participants' overt guessing accuracy did not differ from chance expectations.

More recently, several groups have explored physiological predictors of future events by investigating whether the autonomic nervous system could respond to randomly selected future emotional stimuli.[48,49] Radin designed elegant experiments to evoke an emotional response using randomly selected emotionally arousing or calming photographs, with measures of skin conductance level (SCL) and photoplethysmographic measures of heart rate and blood volume.[39,42] Comparison of SCL response between emotional and calm trials showed a significantly greater change in electrodermal activity around five seconds before a future emotionally arousing picture than before an emotionally neutral or calm picture. These results have since been replicated[43,50-53] and a follow-up study, using functional magnetic resonance imaging, found brain activation in regions near the amygdala *before* emotional pictures were shown, but not before the calm pictures.[53] Consistent findings across previous studies indicate that the body typically responds to a future emotionally arousing stimulus four to seven seconds prior to experiencing the stimulus: however, a later study we conducted suggest this is limited by the experimental protocol. It has also been shown that intuitive perception of a future event is related to the degree of emotional significance one has in the outcome of the event.[43]

Extending and building on Radin's protocol, we added measures of brain response (EEG) and heart rhythm activity (ECG) and found that not only did both the brain and heart receive the pre-stimulus information some 4-5 seconds before a future emotional picture was randomly selected by the computer, but that the heart received this information about 1.3 seconds *before* the brain received it (Figure 2).[10] A number of studies have since found evidence of the heart's role in reflecting future or distant events.[2,54-59] Using a combination of cortical evoked potentials and heartbeat evoked potentials, these studies also found that when the participants were in the heart coherence mode prior to the trials that the afferent input from the heart and cardiovascular system modulated changes in the brain's electrical activity, especially at the frontal areas of the brain. In other words, participants were more attuned to information from the heart when in a coherent state prior to participating in the experimental protocol. Therefore, being in a state of psychophysiological coherence is expected to enhance intuitive ability.[10]

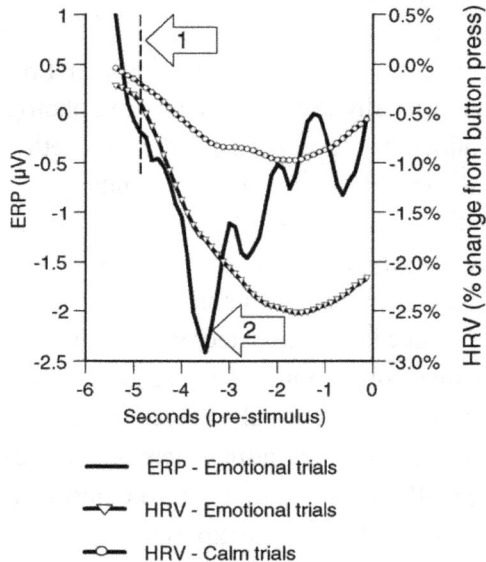

Figure 2 – Example of temporal dynamics of heart and brain pre-stimulus responses: This overlay plot shows the mean event-related potential (ERP) at EEG site FP2 and heart-rate deceleration curves during the pre-stimulus period. (The "0" time point denotes stimulus onset.) The heart-rate deceleration curve for the trials, in which a negative emotionally arousing photo would be seen in the future, diverged from that of the trials that contained a calming future picture (sharp downward shift) about 4.8 seconds prior to the stimulus (arrow 1). The emotional trials ERP showed a sharp positive shift about 3.5 seconds prior to the stimulus (arrow 2). This positive shift in the ERP indicates when the brain "knew" the nature of the future stimulus. The time difference between these two events suggests that the heart received the intuitive information about 1.3 seconds before the brain. Heartbeat-evoked potential analysis confirmed that a different afferent signal was sent by the heart to the brain during this period.[10]

In later studies, we developed a new experimental protocol using a gambling paradigm based on roulette, where the participants are required to choose a bet amount and then choose (bet on) either red or black. A second pre-stimulus segment was added that allowed us to examine the period prior to placing the bet on red or black and the period after the bet was placed but before the randomly determined future outcome (win or lose). In one study using this protocol, we examined individual participants' data over eight separate trials in addition to a group level analysis.[60] Half of the experimental sessions were conducted during the full moon phase and half during the new moon phase in order to assess the potential effects of the moon phase on the pre-stimulus response outcomes and participant winning and amount won ratios. Participants were told that they were participating in a gambling experiment, were given an initial starting kitty, and told they could keep any winnings over the course of 26 trials for each of the eight sessions. The physiological measures included the ECG, from which cardiac inter-beat-intervals (IBI) were derived and skin conductance. We found significant differences between the win and loss responses in the aggregated IBI waveform data during both pre-stimulus segments, which provides important information about nonlocal intuition. On average, we detected a significant pre-stimulus response starting around 18 seconds prior to participants knowing the future outcome. The results of individual participants across the eight sessions found that some, but not all participants, maintained relatively consistent and significant win/loss pre-stimulus response patterns. This is an interesting finding, since there is evidence from other studies that observed effects in repeated trials involving nonlocal information/interaction tended to decline in later trials.[61]

The findings in the analysis of the individual participants lead to several considerations. One is that group level outcomes provide stronger pre-stimulus results. We have previously suggested that measuring the pre-stimulus response in small groups would provide a more stable and reliable outcome indicator for almost any inquiry that can be formulated in a binary format such as red or black, or yes or no, etc.[62] An innovative study that looked at this possibility tested this hypothesis and confirmed that significantly greater pre-stimulus waveform responses are obtained in pairs of participants as compared to single participant pre-stimulus responses.[59] The results of this and other studies also show that if the participants were more attuned to their internal physiological responses, which clearly show a better than chance ability to predict the future outcome than the cognitive level choices, they would have better outcomes in accessing information related to the future.

We also found a significant difference in both pre-stimulus periods during the full moon phase, and amount of money the participant won, but not in the new moon phase. This appears to be consistent with others' findings, for example, Puharich observed increases in the strength of telepathic effects in the 1960s during full moon periods,[63] and Krippner also noticed increased nonlocal perception abilities in the 1970s.[64] More recently, significant solar and full moon phase effects in a large database of psychokinesis experiments were found, and the author suggests that the moon's interaction with the earth's magnetosphere during the moon's passage through the magneto-tail in full moon times may explain the observed effects.[65]

The Roles of the Heart in Accessing Nonlocal Intuition

The research discussed above has shown that when participants shift into a more coherent physiological state prior to engaging in the data collection protocols that a significantly different cardiac afferent (ascending) signal can be detected at the frontal cortex, as well as other brain regions in the pre-stimulus period.[10] It is likely that these signals are important elements of intuition that are particularly salient in pattern recognition and that they are involved in all types of intuitive processes. This implies that learning to attune to and become more conscious of the internal physiological signals, especially the afferent signals from the heart, is an important factor in increasing one's access to nonlocal intuitive information. Given that there is a relationship between increased heart coherence and access to intuitive signals, the capacity to shift into a coherent state is an important factor when considering the heart's role in intuition. Our findings suggest that it's possible to access intuitive intelligence more effectively by first getting into a coherent state, which helps quiet mental chatter and emotional unrest, and then paying attention to shifts in our feelings or perceptions, a process that brings intuitive signals more into conscious awareness.[66] We have also found that increased heart-rhythm coherence correlates with significant improvements in performance on tasks requiring attentional focus and subtle discrimination.[67]

The heart plays a central role in creating coherence and is associated with heartfelt positive emotions. It is not surprising, therefore, that one of the strongest threads uniting the views of diverse cultures and religious and spiritual traditions throughout history has been a universal regard that it is the source of love, wisdom, intuition, courage, etc. Everyone is familiar with such expressions as "put your heart into it," "learn it by heart," "speak from your heart" and "sing with all your heart." All of these indicate that the heart is more than just a physical pump that sustains life. Such expressions reflect what often is called the intuitive, or spiritual heart. Throughout history, people have turned to the intuitive heart, sometimes referred to as their inner voice, soul or higher power, as a source of wisdom and guidance.

We often talk about the "intuitive heart" and "heart intelligence." Both of these terms refer to our *energetic heart*, which we suggest is coupled with a deeper part of the self. Many refer to this as their higher self or higher capacities, or what physicist David Bohm described as our implicate order and undivided wholeness.[4] We use the term "energetic systems" in this context to refer to the functions we cannot directly measure, touch or see, such as our emotions, thoughts and intuitions. Although these functions have loose correlations with biological activity patterns, they nevertheless remain covert and hidden from direct observation. Several notable scientists have proposed that such functions operate primarily in the frequency domain outside of time and space, and have proposed some possible mechanisms that govern how they are able to interact with biological processes.[3,8,68-73]

As discussed elsewhere, the physical heart has extensive afferent connections to the brain and can modulate perception and emotional experience.[67] Our experience suggests that the physical heart also has communication channels connecting it with the energetic heart.[10] An important aspect of nonlocal intuition therefore can be transformational, and from our perspective contains the wisdom that streams from the soul's higher information field down into the psychophysiological system via the energetic heart, and can inform our moment-to-moment experiences and interactions. At the HeartMath Institute, this is what we call "heart intelligence."

Heart intelligence is the flow of higher awareness and the intuition we experience when the mind and emotions are brought into synchronistic alignment with the heart. When we are heart-centered and coherent, we have a tighter coupling and closer alignment with our deeper source of intuitive intelligence. We are able to more intelligently self-regulate our thoughts and emotions, which over time lifts consciousness and establishes a new internal physiological and psychological baseline.[10] In other words, there is an increased flow of intuitive information that is communicated via the emotional energetic system to the mind and brain systems, resulting in a stronger connection with our deeper "inner voice."

There is substantial evidence the heart plays a unique role in synchronizing and stabilizing the activity across multiple systems in the body. As the most powerful and consistent generator of rhythmic information patterns in the body, the heart is in continuous communication with the brain and body through multiple pathways: neurologically through the autonomic nervous system (ANS), biochemically through hormones, biophysically through pressure and sound waves, and energetically through electromagnetic-field interactions. Because of these multiple communication pathway, the heart is uniquely positioned to act as the "global coordinator" in the body's symphony of functions to synchronize the system as a whole.[33,67,74] Because of the extensiveness of the heart's influence on physiological, cognitive and emotional systems, the heart provides a central point from which the dynamics of the psychophysiological systems can be self-regulated.

One of the research focuses of our laboratory over the last decade has been the study of the patterns and rhythms generated in various physiological systems during the experience of different thoughts, emotions, and behaviors. By experimenting with numerous physiological measures, we found that heart rate variability (heart-rhythm) patterns are consistently dynamic and reflective of changes in one's emotional state (Figure 3).[67,75]

It is important to note that although heart rate often changes with emotions, our research has found that it is the pattern of the heart's rhythm that is primarily reflective of the emotional state.[67,76,77] Spanish researcher Enrique Leon expanded on our observations by analyzing the

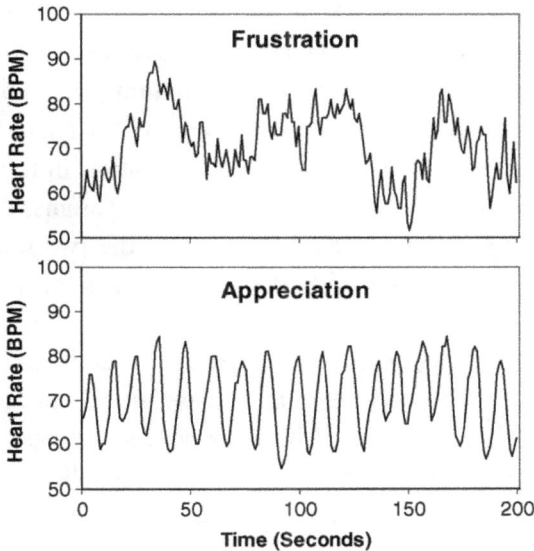

Figure 3 – Emotions are reflected in heart-rhythm patterns. The heart-rhythm patterns shown in the top graph, characterized by its erratic, irregular pattern (incoherence) is typical of depleting emotions such as anger or frustration. The bottom graph shows an example of the coherent heart-rhythm pattern that is typically observed when an individual is experiencing sustained, modulated regenerative emotions, in this case appreciation. Both recordings are from the same individual only a couple of minutes apart. The amount of variability and mean heart rate are the same in both examples, illustrating how the pattern of activity contains information in the absence of changes in physiological activation.

rhythmic patterns that occur in heart rate variability (HRV). Leon found that by analyzing HRV patterns, as measured by a heart-rhythm monitor, he could correctly identify discrete emotional states such as anxiety versus frustration with 75% accuracy.[78] These changes in rhythmic heart-rhythm patterns can be independent of heart rate, meaning that one can have varying degrees of coherent or incoherent patterns at higher or lower heart rates. Thus, it is the pattern of the rhythm rather than the rate at any point in time that is most directly related to emotional dynamics and physiological synchronization.[67] This is important because the heart-rhythm pattern that is generated, regardless if it is incoherent or coherent, affects brain centers involved in sensory motor integration, decision-making, problem-solving, self-regulation and behavior. It also is related to our ability to access our intuition.

Physiological coherence, which also can be referred to as heart coherence, cardiac coherence, or resonance, is be assessed by analyzing HRV rhythms in specific ways. A person's heart-rhythm pattern becomes more ordered and sine-wavelike at a frequency around 0.1 Hz (10 seconds).[67,75] A coherent state reflects increased synchronization in the activity and flow of information between higher-level brain systems, more efficient activity occurring between the interactions of the two branches of the ANS (sympathetic and parasympathetic), and an increase in parasympathetic activity, or what often is called vagal tone.[67] Importantly, there is physiological evidence that the continued practice of coherence-building techniques creates a *repatterning process* in the neural architecture, where coherence becomes established as a new, stable baseline reference memory.[79] In a practical sense, coherence becomes the new set point or default, facilitating access to intuition and the ability to self-regulate emotions and stress responses, a process that then becomes increasingly familiar and eventually automatic.[80,81,82,83] This makes it easier for people to maintain their "center" and increase their mental and emotional flexibility and remain in self-directed control. It also builds one's capacity to access all three types of intuition.

Benefits of Intuition

Access to our heart's intuition varies among people, but we all have it. As we learn to slow down our minds and attune to our deeper heart feelings, a natural intuitive connection can occur. Intuition often is thought of in the context of inventing a new light bulb or winning in Las Vegas, but what most people discover is that is intuition is a very practical asset that can help guide their moment-to-moment choices and decisions in daily life. Our intuitive insights often unfold better understanding of ourselves, others, issues, and life than years of accumulated knowledge. It is especially helpful for eliminating unnecessary energy expenditures, which deplete our internal reserves, making it more difficult to self-regulate and be in charge of our attitudes, emotions, and behaviors in ordinary day-to-day life situations. Intuition allows us to increase our ability to move beyond automatic reactions and perceptions. It helps us make more intelligent decisions from a deeper source of wisdom, intelligence, and balanced discernment, in essence elevating consciousness, happiness, and the quality of our life experience. This increases synchronicities and enhances our creativity and ability to flow through life. It also increases our ability to handle awkward situations, such as dealing with difficult people, with more ease and promotes harmonious interaction and connectivity with others.

One of the most important keys to accessing more of our intuitive intelligence and inner sense of knowing is developing deeper levels of self-awareness of our more subtle feelings and perceptions, which otherwise never rise to conscious awareness. In other words, we have to pay attention to the intuitive signals that often are under the radar of conscious perception or are drowned out by ongoing mental chatter and emotional unrest. A common report from people who practice being more self-aware of their inner signals is that the heart communicates a steady stream of intuitive information to the mind and brain. In many cases, however, we only perceive a small percentage of intuitive information or choose to override the signals because they do not match our more ego-centric desires.

The HeartMath System of self-regulation techniques was informed by research on heart-brain interactions and optimal function.[84-88] The HeartMath System offers people practical and reliable techniques for increasing physiological coherence and self-regulating from a state of emotional unease or stress into a "new" positive state of emotional calm and greater stability. Studies have been conducted across diverse populations in laboratory, organizational, educational, and clinical settings on HeartMath coherence-building techniques. The studies have shown these techniques are effective in producing immediate and sustained reductions in stress and its associated disruptive and dysfunctional emotions, and improving many dimensions of health and well-being.[67,74,83,89-92] Collectively, results indicate such techniques are easily learned and employed, produce rapid improvements, have a high rate of compliance, can be sustained over time and are readily adaptable to a wide range of ages and demographic groups.

Conclusion

This chapter has explored different perspectives about the nature and types of intuition and the connection between intuition and lifting consciousness. We have suggested that increased effectiveness in self-regulatory capacity and the resultant reorganization of implicit memories sustained in the neural architecture facilitates a stable and integrated experience of self and our relationship to others and the larger environment. We have also indicated that there are many benefits to be gained by a deeper understanding of the complex interactions between the heart, brain, and the energetic heart. Learning to access our deeper innate wisdom can help people unfold who they really are and approach personal, social, and global affairs with more wisdom, compassion, and positive innovation.

When we practice shifting to a more coherent state it increases intuitive awareness and, over time, the establishment of new baseline reference patterns and sustained shifts in perception and world-views. We can discern more informed and intelligent decisions from these new patterns and shifts in perception. This process elevates our consciousness, awareness of self, and connections with others, as well as the capacity for self-regulation and the corresponding ability for self-directed action.

As the development of physiological coherence allows increased access to intuitive intelligence and one's repertoire of positive emotions and actions grows, it is natural that the enhanced experience of empathy and social coherence will lead to compassionate actions and behaviors that promote and support altruistic pro-social behaviors. When more individuals in families, workplaces, and communities increase and stabilize their coherence baselines, it can lead to increased social and global coherence and a corresponding lifting of overall human consciousness.

Endnotes

1 Myers, D. G. *Intuition: Its Powers and Perils.* (Yale University Press, 2002).
2 Bradley, R. T., R. McCraty, M. Atkinson, & M. Gillin. in *Regional Frontiers of Entrepreneurship Research.*
3 Bradley, R. T. Psycholphysiology of Intution: A quantum-holgraphic theory on nonlocal communication. *World Futures: The Journal of General Evolution* 63, 61–97 (2007).
4 Bohm, D. & Hiley, B. J. *The Undivided Universe.* (Routledge, 1993).

5 Laszlo, E. *The Interconnected Universe: Conceptual Foundations of Transdiciplinary Unified Theory.* (World Scientific, 1995).

6 Loye, D. *The Sphinx and the Rainbow: Brain, Mind and Future Vision.* (Bantam Books, 1983).

7 Nadeau, R. & Kafatos, M. *The Nonlocal Universe: The New Physics and Matters of the Mind.* (Oxford University Press, 1999).

8 Pribram, K. H. *Brain and Perception: Holonomy and Structure in Figural Processing.* (Lawrence Erlbaum Associates, Publishers, 1991).

9 Sheldrake, R., McKenna, T. & Abraham, R. *The Evolutionary Mind: Trialogues at the Edge of the Unthinkable.* (Trialogue Press, 1998).

10 McCraty, R., Atkinson, M. & Bradley, R. T. Electrophysiological evidence of intuition: Part 2. A system-wide process? *Journal of Alternative and Complementary Medicine* 10, 325-336 (2004).

11 McCraty, R. & Zayas, M. The Heart's Role in Accessing Intuitive Intelligence and Lifting Consciousness. *Global Advances in Health and Medicine* (2014).

12 Lieberman, M. D. in *Annual Review of Psychology* Vol. 58 *Annual Review of Psychology* 259–289 (Annual Reviews, 2007).

13 Hodgkinson, G. P., Langan-Fox, J. & Sadler-Smith, E. Intuition: A fundamental bridging construct in the behavioural sciences. *British Journal of Psychology* 99, 1–27 (2008).

14 Dane, E. & Pratt, M. G. Exploring intuition and its role in managerial decision making. *Academy of Management Review* 32, 33–54 (2007).

15 Bechara, A., Damasio, H., Tranel, D. & Damasio, A. R. The Iowa Gambling Task and the somatic marker hypothesis: some questions and answers. *Trends in cognitive sciences* 9, 159–162; discussion 162-154, doi:10.1016/j.tics.2005.02.002 (2005).

16 Damasio, A. R. *Descartes' Error: Emotion, Reason and the Human Brain.* (G.P. Putnam's Sons, 1994).

17 Bastick, T. *Intuition: How we think and act.* (Wiley, 1982).

18 Moir, A. & Jessel, D. *Brainsex: The real difference between men and women.* (Mandarin Paperbacks, 1989).

19 Agor, W. *Intuitive Management: Integrating Left and Right Brain Skills.* (Prentice Hall, 1984).

20 Eisenhardt, K. & Zbaracki, M. Strategic decision making. *Strategic Management Journal* 13, 17–37 (1992).

21 Laughlin, C. in *Intuition: The Inside Story* (eds R Davis-Floyd & P S Arvidson) 19–37 (Routledge, 1997).

22 Hogarth, R. M. *Educating Intuition.* (The University of Chicago Press, 2001).

23 Torff, B. & Sternberg, R. J. in Understanding and Teaching the Intuitive Mind: Student and Teacher Learning (eds B Torff & R J Sternberg) 3–26 (Lawrence Erlbaum Associates, Publishers, 2001).

24 Larsen, A. & Bundesen, C. A template-matching pandemonium recognizes unconstrained handwritten characters with high accuracy. *Memory & cognition* 24, 136–143 (1996).

25 Craig, J. & Lindsay, N. Quantifying "gut feeling" in the opportunity recognition process. *Frontiers of Entrepreneurship Research*, 124–135 (2001).

26 Mayer, R. E. in *The nature of insight* (eds R. J. Sternberg & J. E. Davidson) 3–32 (The MIT Press, 1996).

27 McCraty, R. in *Bioelectromagnetic and Subtle Energy Medicine, Second Edition* (ed Paul J. Rosch) (2015).

28 Hammerschlag, R. *et al.* biofield Physiology: A Framework for an emerging discipline. *Global Advances in Health and Medicine* 4, 35–41 (2015).

29 Halberg, F., Cornelissen, G., McCraty, R. & A.Al-Abdulgader, A. Time Structures (Chronomes) of the Blood Circulation, Populations' Health, Human Affairs and Space Weather. *World Heart Journal* 3, 1–40 (2011).

30 Alabdulgader, A. et al. Human heart rhythm sensitivity to earth local magnetic field fluctuations. *Journal of Vibroengineering* 17 (2015).

31 Uyeda, S., Nagao, T., Orihara, Y., Yamaguchi, T. & Takahashi, I. Geoelectric potential changes: possible precursors to earthquakes in Japan. *Proc Natl Acad Sci U S A* 97, 4561-4566 (2000).

32 Wiseman, R. & Schlitz, M. Experimenter effects and the remote detection of staring. *Journal of Parapsychology* 61, 197–207 (1997).

33 McCraty, R. in *Bioelectromagnetic Medicine* (eds P J Rosch & M S Markov) 541–562 (Marcel Dekker, 2004).

34 Hatfield, E. Emotional Contagion. (Cambridge University Press, 1994).

35 Konvalinka, I. & Roepstorff, A. The two-brain approach: how can mutually interacting brains teach us something about social interaction? *Frontiers in human neuroscience* 6, 215, doi:10.3389/fnhum.2012.00215 (2012).

36 Morris, S. M. Facilitating collective coherence: Group Effects on Heart Rate Variability Coherence and Heart Rhythm Synchronization. *Alternative Therapies in Health and Medicine* 16, 62–72 (2010).

37 Jahn, R. G. & Dunne, B. J. *Margins of reality: The role of consciousness in the physical world.* ICRL Press, 2009).

38 Bradley, R. T., Gillin, M., McCraty, R. & Atkinson, M. Nonlocal Intuition in Entrepreneurs and Non-entrepreneurs: Results of Two Experiments Using Electrophysiological Measures. *International Journal of Entrepreneurship and Small Business* 12, 343–372 (2011).

39 Radin, D. *The Conscious Universe: The Scientific Truth of Psychic Phenomena.* (HarperEdge, 1997).

40 Bem, D. J. Feeling the future: experimental evidence for anomalous retroactive influences on cognition and affect. *J Pers Soc Psychol* 100, 407–425, doi:10.1037/a0021524 (2011).

41 Mossbridge, J., Tressoldi, P., E & Utts, J. Predictive Physiological Anticipation Preceding Seemingly Unpredictable Stimuli: A Meta-Analysis. *Frontiers in psychology* 3:390 (2012).

42 Radin, D. I. Unconscious perception of future emotions: An experiment in presentiment. *Journal of Scientific Exploration* 11, 163–180 (1997).

43 McCraty, R., Atkinson, M. & Bradley, R. T. Electrophysiological evidence of intuition: Part 1. The surprising role of the heart. *Journal of Alternative and Complementary Medicine* 10, 133–143 (2004).

44 Levin J, K. J. The relationship of slow cortical potentials to psi information in man. *J Parapsychol* 39, 25–26 (1975).

45 Warren CA, M. B., Don NS.169-181. in *The Parapsychological Association 35th Annual Convention: Proceedings of Presented Papers.* 169–181

46 Don, N. S., McDonough, B. E. & Warren, C. A. Event-related brain potential (ERP) indicators of unconscious psi: A replication using subjects unselected for psi. *Journal of Parapsychology* 62, 127–145 (1998).

47 McDonough, B. E., Don, N. S. & Warren, C. A. Differential event-related potentials to targets and decos in a guessing task. *Journal of Scientific Exploration* 16, 187–206 (2002).

48 Spottiswood, J. & May, E. Skin conductance prestimulus response: Analyses, artifacts and a pilot study. *Journal of Scientific Exploration* in press (2003).

49 Radin, D. May et al.'s "anomalous anticipatory skin conductance response to acoustic stimuli". *J Altern Complement Med* 11, 587–588, doi:10.1089/acm.2005.11.587 (2005).

50 Bem, D. J. Precognitive habituation: Replicable evidence for a process of anomalous cognition. *Unpublished manuscript* (2003).

51 Bierman, D. J. & Radin, D. I. Anomalous anticipatory response on randomized future conditions. *Percept Mot Skills* 84, 689–690 (1997).
52 Bierman, D. J. in *Proceedings of Presented Papers: The 43rd Annual Convention of the Parapsychological Association*. 34–47.
53 Bierman, D. J. & Scholte, H. S. in *Presented at Towards a Science of Consciousness* IV.
54 Tressoldi, P. E., Martinelli, M., Massaccesi, S. & Sartori, L. Heart rate differences between targets and non targets in intuition tasks. *Fiziologiia cheloveka* 31, 32–36 (2005).
55 Hu, H. & Wu, M. New Nonlocal Biological Effect. *NeuroQuantology* 10, 462–467 (2012).
56 Tressoldi, P. E., Massimiliano, M., Zaccaria, E. & Massaccesi, S. Implicit Intuition: How Heart Rate can Contribute to Prediction of Future Events. *Journal of the Society for Psychical research* 73, 1–16 (2009).
57 Sartori, L., Massacessi, S., Martinelli, M. & Tressoldi, P. E. Physiological correlates of ESP: heart rate differences between targets and nontargets. *Journal of Parapsychology* 68, 351 (2004).
58 Tressoldi, P. E., Martinelli, M., Scartezzini, L. & Massaccesi, S. Further evidence of the possibility of exploiting anticipatory physiological signals to assist implicit intuition of random events. *Journal of Scientific Exploration* 24, 411 (2010).
59 Toroghi, S. R., Mirzaei, M., Zali, M. R. & Bradley, R. T. Nonlocal Intuition: Replication and Paired-Subjects Enhancement Effects. *Global Advances in Health and Medicne* (2014).
60 McCraty, R. & Atkinson, M. Electrophysiology of Intuition: Pre-stimulus Responses in Group and Individual Participants Using a Roulette Paradigm. *Global advances in health and medicine: improving healthcare outcomes worldwide* 3, 16-27, doi:10.7453/gahmj.2014.014 (2014).
61 ATMANSPACHER, H. & JAHN, R. G. Problems of Reproducibility in Complex Mind-Matter Systems. *Journal of Scientic Exploration* 17, 243–270 (2003).
62 R. M., Atkinson, M. & Childre, D. Electrophysiological Intuition Indicator US patent (2005).
63 Puharich, A. *Beyond Telepathy*. (Anchor Books 1973).
64 Krippner, S., Becker, A., Cavallo, M. & Wahhburn, B. Electrophysiological studies of ESP in dreams: Lunar cycle differences in 80 telepathy sessions., 14–19 (Human Dimensions Institute Bufflao, NY, 1972).
65 Eckhard, E. Solar-Periodic Full Moon Effect in the Fourmilab RetroPsychoKinesis Project Experiment Data: An Exploratory Study. *The Journal of Parapsychology* 69, 233–261 (2005).
66 Petitmengin-Peugeot, C. in *The View from Within*. First-person approaches to the study of consciousness (eds F.J.Varela & J. Shear) 43–77 (Imprint Academic, 199).
67 McCraty, R., Atkinson, M., Tomasino, D., & Bradley, R. T. The coherent heart: Heart-brain interactions, psychophysiological coherence, and the emergence of system-wide order. *Integral Review* 5, 10–115 (2009).
68 Laszlo, E. Quantum Shift in the Global Brain: how the new scientific reality can change us and our world (Inner Traditions, 2008).
69 Mitchell, E. in *Bioelectromagnetic Medicine* (eds P.G. Rosch & M.S. Markov) 153–158 (Dekker, 2004).
70 Tiller, W. A., W E Dibble, J. & Kohane, M. J. *Conscious Acts of Creation: The Emergence of a New Physics*. (Pavior Publishing, 2001).
71 Marcer, P. & Schempp, W. The brain as a conscious system. *Internationl Journal of General Systems* 27, 231–248 (1998).
72 Pribram, K. H. & Bradley, R. T. in *Self-Awareness: Its Nature and Development* (eds M Ferrari & R Sternberg) 273–307 (The Guilford Press, 1998).

73 Schempp, W. Quantum holograhy and neurocomputer architectures. *Journal of Mathematical Imaging and vision* 2, 109–164 (1992).

74 McCraty, R., Childre, D. Coherence: Bridging Personal, Social and Global Health. *Alternative Therapies in Health and Medicine* 16, 10–24 (2010).

75 McCraty, R., Atkinson, M., Tiller, W. A., Rein, G. & Watkins, A. The effects of emotions on short term heart rate variability using power spectrum analysis. *American Journal of Cardiology* 76, 1089–1093 (1995).

76 McCraty, R., Atkinson, M., Tiller, W. A., Rein, G. & Watkins, A. D. The effects of emotions on short-term power spectrum analysis of heart rate variability. *American Journal of Cardiology* 76, 1089–1093 (1995).

77 Tiller, W. A., McCraty, R. & Atkinson, M. Cardiac coherence: A new, noninvasive measure of autonomic nervous system order. *Alternative Therapies in Health and Medicine* 2, 52–65 (1996).

78 Leon, E., Clarke, G., Callaghan, V. & Dotor, F. Affect-aware behavior modelling and control inside an intelligent environment *Pervasive and Mobile Computing doi:10.1016/j. pmcj.2009.12.002* doi:10.1016/j.pmcj.2009.12.002 (2010).

79 Bradley, R. T., McCraty, R., Atkinson, M., Tomasino., D. Emotion Self-Regulation, Psychophysiological Coherence, and Test Anxiety: Results from an Experiment Using Electrophysiological Measures. *Applied Psychophysiology and Biofeedback* 35, 261–283 (2010).

80 McCraty, R., Atkinson, M., Tomasino, D. & Bradley, R. T. *The coherent heart: Heart-brain interactions, psychophysiological coherence, and the emergence of system-wide order.* (HeartMath Research Center, Institute of HeartMath, Publication No. 06–022, 2006).

81 McCraty, R. & Childre, D. in *The Psychology of Gratitude* (eds R A Emmons & M E McCullough) 230–255 (Oxford University Press, 2004).

82 McCraty, R., Atkinson, M. & Tomasino, D. Impact of a workplace stress reduction program on blood pressure and emotional health in hypertensive employees. *Journal of Alternative and Complementary Medicine* 9, 355–369 (2003).

83 Alabdulgader, A. Coherence: A Novel Nonpharmacological Modality for Lowering Blood Pressure in Hypertensive Patients. *Global Advances in Health and Medicne* 1, 54–62 (2012).

84 Childre, D. L. *Freeze-Frame®, Fast Action Stress Relief.* (Planetary Publications, 1994).

85 Childre, D. & Martin, H. *The HeartMath Solution.* (HarperSanFrancisco, 1999).

86 Childre, D. & Cryer, B. *From Chaos to Coherence: The Power to Change Performance.* (Planetary, 2000).

87 Childre, D. & Rozman, D. *Overcoming Emotional Chaos: Eliminate Anxiety, Lift Depression and Create Security in Your Life.* (Jodere Group, 2002).

88 Childre, D. & Rozman, D. *Transforming Stress: The HeartMath Solution to Relieving Worry, Fatigue, and Tension.* (New Harbinger Publications, 2005).

89 McCraty, R., Atkinson, M. & Tomasino, D. Impact of a workplace stress reduction program on blood pressure and emotional health in hypertensive employees. *J Altern Complement Med* 9, 355–369, doi:10.1089/107555303765551589 (2003).

90 Ginsberg, J. P., Berry, M.E., Powell, D.A. Cardiac Coherence and PTSD in Combat Veterans. *Alternative Therapies in Health and Medicine* 16, 52–60 (2010).

91 Lloyd, A., Brett, D., Wesnes, K. Coherence Training Improves Cognitive Functions and Behavior In Children with ADHD. *Alternative Therapies in Health and Medicine* 16, 34–42 (2010).

92 Bedell, W. Coherence and hearlth care cost—RCA acturial study: A cost-effectivness cohort study *Alternative Therapies in Health and Medicine* 16, 26–31 (2010).

Live in Spirit

Steve Curtis

The first peace, which is the most important, is that which comes within the souls of people when they realize their relationship, their oneness, with the universe and all its powers, and when they realize that at the center of the universe dwells the Great Spirit, and that its center is really everywhere, it is within each of us.
—Nicholas Black Elk

It was no longer white. And it wasn't covered in just a few simple lines, as it had been on other days. It faced me from the side of the room as class went on and I just kept looking at it in awe: a living image for how much more there was to life if I let it in, how much more there was to learn and discover.

Every day for the last week there was a blank sheet of paper on the wall, and as participants entered the room, we were encouraged to make one mark on it: lines, swirls, colors, and shapes, but no words or symbols, since those were constructs of the left brain. Every day something different had appeared, and I would add my mark to what was already there. I had wanted to be the first one to mark up the paper on one of the days, but I tend to take my time in the morning and was always too late. For the past few days, though, with increasing commitment, I had resolved to get up just a little bit earlier so that I would be first. In fact, I was so sure this would happen on one particular morning that I planned what I was going to draw the day before. I woke up with it vivid in my mind.

My experience over the past few weeks had been transformative, and I wanted to express that with an image of an ocean with a blue bottom and blue top and red bubbles coming up from below. To me this symbolized the wisdom I felt emerging from within. It was very clear and felt incredibly intense to me. I had to draw it. Sure, I'd be taking more than my fair share of the canvas, and I knew that symbols weren't in the protocol, but I couldn't resist.

I arrived at the center that morning an hour early, eagerly looking forward to making my mark. When I got there, I was dumbfounded. Not only had someone beaten me to the canvas, he or she had drawn exactly what I had planned. Exactly! Same colors, same pattern, same place on the paper. It was just as I had envisioned it. How could this be? Did I tell someone? No. Was it just a coincidence? No, it was too specific and nothing in previous days looked even close to this. Who drew this, I wondered? After a few minutes of investigating, I discovered that it had come from more than one person. Each had made their mark to create the vision I'd had in my mind. If it were any other weekday, I might have dismissed it as a one-in-a-million anomaly, an extreme coincidence. But this was no average weekday. It was the second week of Stephen Gilligan's Trance Camp, and nothing about Gilligan or his classes is average.

Gilligan is one of the world's foremost experts on altered states of consciousness,[1] and specifically on what's called "generative trance." It's similar to NLP (neuro-linguistic programming, a technique that uses language, behavior, and the nervous system to reprogram thought patterns) but much more powerful. Gilligan has all the power of a stage hypnotist, but instead of doing tricks, he surrenders to the wisdom of spirit and the subconscious mind. He describes trance as a naturally flowing process that cannot be forced; it can only emerge organically, without effort. He believes that this altered state can be used for anything from enhancing creativity and overcoming phobias to learning to love again after traumatic loss. Just about anything that the mind can conceive is possible by suspending belief and entering into the full field of possibility that trance offers.

Gilligan is an amazing man, with a presence unlike anyone I have ever met. To be in his space for just a few minutes transforms you and what you think you are capable of. It's hard to describe in words. Just being around him seems to release you from the stress and rigidity of the moment and into a place of utter calm and freedom where the rules we normally live by, that tell us what we can or can't do or how we can or can't feel, cease to exist. It's a powerful and seductive experience. For all of Gilligan's talents, though, he says none of them compare to those of his teacher, Milton Erikson, the father of modern hypnosis. Erikson's understanding of the trance state was so profound that he could bring a person into one simply from eye contact or by a light touch of his hand. According to Gilligan,[2] who aspires to the same ability, Erikson was an absolute master whose own experiences were a testament to the power of the unconscious mind.

At age 17, Erikson contracted polio and was so severely paralyzed that doctors believed he would die. Using autohypnotic (self-induced) suggestions,[3] he not only survived, but slowly regained control of parts of his body and was eventually able to talk and use his arms. Still unable to walk, though, he decided to train his body further by embarking, alone, on a one-thousand-mile canoe trip. After this grueling journey, he was able to walk with a cane. In essence, he organized his body to do something that was considered impossible by conventional medicine. This is exactly why I was here. My diagnosis had set very clear and very rigid limits, and I wanted to find people who had not only pushed their limits but smashed through them. Erikson was one of these people.

My experience with the wall drawing was just one of many phenomenal things that occurred during those two weeks. Time and time again as I let go of my mind and the constraints I held, such as the many details I normally obsessed over or the limiting beliefs that governed my life, magic began to emerge. Here's another example: I was walking back from lunch with some friends. We saw some fellow classmates up ahead and I suddenly had a vision of one of them saying, "I am free, like an eagle," and moving his arms back and forth like a bird in flight.

As I passed one of the students, I said, "Hi, how are you?" He said, "I am free, like an eagle," and started waving his arms exactly as I had envisioned it. Synchronicity is a normal part of life, but this was so specific, and it had happened so many times over the past week, that it couldn't be the only explanation. Something else was happening.

A Land Far Away, in Your Own Backyard

The state of mind we ultimately enter at Trance Camp is far from the ordinary world; it opens a space where such non-ordinary experiences are common. It's as if the past, the present, and the future all become one in a dream state that has no boundaries. As I write this, I am compelled to admit that while I'd always considered myself a skeptic and a realist, the experiences I had there rearranged the way I was seeing myself as well as the world.

Gilligan made it clear from the start that within us is a profound force of consciousness that possesses the wisdom of many millennia. It's a force of immense potential, with access to all the information that is or ever was—a living library of human experience (including our own) and otherworldly knowledge. Since I was most concerned about healing my terminal illness, I asked many questions about the potential of this force and the power of trance to heal a serious condition. Gilligan's answers, however, were always frustratingly vague.

What *was* becoming clearer, though, is how our conscious minds often get in the way of the healing process. Turn off the noise and other channels open up and become available—channels that sometimes run deep below the surface. Techniques such as trance, NLP, and hypnosis essentially by-pass our fixed, day-to-day mental states in order to access our subconscious. They are being used to treat a variety of conditions, from unhealthy habits[4] to phobias[5] and addictions, but no one has developed a model of their use to cure the terminally ill—though there have been some very intriguing cases. One involved a man named Robert Dilts, a leader and pioneer in the field of NLP for the last three decades. His mother, Pat, had a history of recurring, inoperable breast cancer, and in 1984 Dilts spend four consecutive days with her, working to change every thinking pattern that may be contributing to her condition or her healing. In the days and weeks after the session, her cancer went into remission without further use of chemo, radiation, or other conventional treatment and she lived another twelve years, confounding the expectations of her doctors. Dilts wrote about this remarkable experience in a 1990 book, *Beliefs: Pathways to Health and Well-Being*[6] (a revised second edition was published in 2012).

There have since been other documented[7] cases of NLP's role in cancer remission, and while Dilts himself never claimed that NLP alone can heal people with serious illnesses, the fact that our beliefs might play a central role in the healing process was incredibly exciting to me. I was also sobered: I knew that if I was going to find a solution to my situation, I would need to be VERY aware and focused.

Gilligan described illness as a symptom of the pressure caused by the diverging objectives of the subconscious and conscious mind. Like two tectonic plates that push against each other, the pressure builds and slowly starts causing mini tremors that appear as symptoms in the body. If the pressure and friction continue to build, a massive earthquake occurs that can threaten the existence of the organism itself—just as an earthquake in Los Angeles would bring the entire city to a grinding halt.

His description made it clear that the alignment between mind and spirit were of utmost importance, so I started asking myself, how closely *my* mind and spirit were aligned? What were their objectives? How could I hear what each side wanted?

During the second week of Trance Camp we organized into teams of three and were instructed to use the technique we'd been taught to lead each other into trance. The goal was to create an outcome that was meaningful to our life. Each person on the team would take turns being the 'subject,' while the other two acted in the roles of leader and assistant. I was the first subject for my team.

We took our chairs out into a warm field near the hotel. The leader asked me to close my eyes and did some basic hypnotic induction, like encouraging me to focus on my breath to help me relax. Once I settled into trance, a state I recognized by a profound sense of relaxation, a near empty mind, and a warm fuzzy feeling all over my body, she asked me to choose an intention for my session. Like in the bubbles in the drawing, the words appeared in my mind seemingly out of nowhere.

"I would like my creativity to flow more easily," I said, barely realizing that I had spoken aloud.

I was so self-conscious about being judged that I had trouble speaking in public, let alone expressing creative thoughts through poetry, singing, or dance. I had been in sales for many years, and I had no problem talking to people or negotiating with them, but this was easy because I didn't need to show who I was. Showing the real me—my feelings, my dreams, my vulnerabilities, my pain—was a very different thing. After all, isn't that what makes performing artists so mesmerizing or leaders so charismatic? I believe it's the depth and degree to which they are able to accept, be with, and experience their humanity in the presence of others. I had a deep sense of knowing I could get there. If I could just get past whatever was blocking me from being real in the presence of others, something very powerful would emerge.

The leader then asked me to name the "resources" in my life—images, memories of people, sounds, places, or things—that could help me to be more creative. I shared some experiences that gave me confidence: my grandfather smiling at me in way that said I could never do anything wrong; my mother telling me "You can do it!" with complete faith that I really could; the confidence of a few achievements like winning a race with a 4:20 mile or my first parachute jump during special forces training. We then started the processing part of the trance.

The leader led me back in time, encouraging me to relax until I was in a place of light and all words became very soft. A team member was sitting beside me, playing the role of assistant. His job was to facilitate the trance by making soothing, nondescript sounds that helped maintain my relaxed state but also confused my conscious mind so it would let go.

"Weee wwwwasssssshhhhh click clik yeeeeaaaaa ha ha ha click shh-hhh sshhhhhhh shhhhhsssssssss," I heard over and over in the back-ground. The first time I heard this a week ago I couldn't stop laughing. But I now knew how potent this simple tool was in prying the mind loose from its death grip on our highest expression of ecstasy—true presence.

As the leader guided me to a place of greater creativity, she inter-spersed stories about supportive people in my life with various comments about the new way of being I was seeking: "You feel your creativity swelling up from a deep well." "Your grandfather is watching, he can see how deep it is." "It is coming forth with a power so great that it can-not be stopped." "Your mother knows this will makes anything possible for you." And so on. She followed all this with a closing sequence that brought me back out of trance.

For most of this time I wasn't thinking about anything. I was just experiencing the emotions, feelings, and images that came up without thoughts, beliefs or words. It was as Gilligan had described it, "stirring the pot" with my intention, along with doses of new beliefs and powerful reinforcements of personal strength and freedom. While it felt like only two minutes had passed, it had been 20. Afterwards something was dif-ferent. My words flowed in a way that they didn't before. I was light, care-free, and inspired. The timing was perfect, as there was a poetry session later. While I had avoided the readings so far, now I wanted to go.

That night I did something I had never done: I wrote a poem, read it for the group, and everyone loved it! To this day I still write poetry as a way of expressing and releasing my emotions. Such a process had been so far from my reality. How did my long-held block evaporate in a mere 20 minutes? How could it now be so effortless? What else could I do using this technique? In the coming years, I would find out.

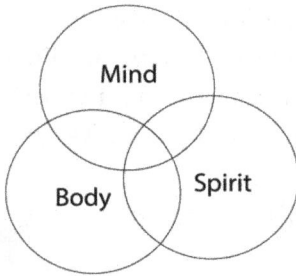

The conflict, pressure, and friction that Gilligan spoke of between the conscious and unconscious mind was something I was very aware of. I felt trapped in my business and my life. I wasn't happy with what I was doing or how I was doing it. Everything seemed out of balance.

The Holy Trinity

As my healing journey evolved, I came to see it more and more in the same way I see it today, as a divine partnership. The foundation of health, peace, and joy depends on the balanced relationship between the mind, the body, and the spirit—each of them equal partners. For me, each has its own objective in managing the whole: The mind's purpose is to maintain the ego and filter information; to gather, achieve, protect, and replicate. The spirit's purpose is to reach outward to the "We": to maintain balance and harmony with all other life, to guide, inspire, and shepherd. The body's purpose is to give both the mind and the spirit a home, a place of support and sustenance. Optimal living—life satisfaction, quality of life, and well-being—comes from honoring the role of each[8] and also recognizing that their work is tightly intertwined.

If the body leads alone, one will have a meaningless, primal, and short-lived existence. If the mind leads alone, a person can become isolated, delusional, and destructive. Even spirit alone is not sufficient—it needs the other two for balance, to help guide this journey we call life.

When one force takes over and excludes the other viewpoints, the neglected two will push back and team up to get its attention. Like most partnership disputes (families, businesses teams, etc.), there are some initial conversations, but if no one listens, the conflict escalates. In business, employees start acting out, factions form, groups fight and turn teams against teams, lawsuits might be filed, and the conflict spreads through the entire system, threatening and possibly destroying the whole enterprise. The greater the gap and the mistrust, the more difficult it is to resolve the disagreement. Working things out requires time, patience, openness, honesty, and consistency—as well as mutual commitment.

While business partners can always part ways, this isn't an option for the fundamental relationship of body-mind-spirit that makes up our life. They are equal parts of a whole. When the three are in harmony, life feels effortless—filled with joy and full of vitality. This is what I experience most every day, and it couldn't be more different from how I was prior to receiving my diagnosis.

Flow from Source

The subject of *being* is a complicated one to describe, but I will try, using a metaphor from the world of geology. Imagine the flow of consciousness into the world as magma flows into a volcano: they both come from some deep, unseen place and then visibly manifest in a multitude of ways. On earth, molten lava circulates in the planet's core as an undifferentiated whole. It moves through the various layers of a volcano before bursting into visibility.

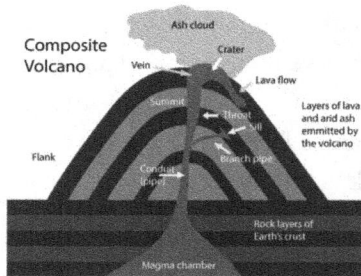

Sometimes the magma flows smoothly, letting off pressure as it rises up from its source. If too much pressure builds up in any part of the system, though, the result can be explosive. Think of the modest but spectacular eruptions of Mount Kilauea in Hawaii compared to the emotive, destructive force of Mount St. Helens.

Like the birth of a volcano, human beings and the magma of human consciousness emerge from a vast, unseen source, to be mysteriously shaped and individualized by one's unique calling or destiny, by the environment they grow up in, the genes they inherited, the stories they are told about how reality works, and how it all manifests in their developing minds. Even as adults, our thoughts and emotions are influenced by the magma of belief systems that stir and swirl beneath our conscious awareness.

Although there is little agreement among established theologies regarding the ultimate nature of God – from the personal god of Christianity to Buddhism's fertile void—it's all still a mystery, our experience of it filtered through our conditioning and beliefs. Huston Smith, who wrote *The World's Religions*[9] in 1958, which is still widely used as a textbook on comparative religion, concluded that the main thing common to all religious belief systems is treating others well. He also said they all have rules, and they all have mystical traditions that recognize divine experience—an experience of oneness, of connection with all—as accessible to anyone without the need for dogma. And now we have quantum physics telling us that matter is basically an illusion, that everything is energy, and it's all interconnected (more on that later in the chapter). This is where it really gets interesting.

I believe that this interconnectedness has something to do with an ultimate Source, and that when we block this Source, we block our ability to heal and reach our greatest potential. The thing that clogs it up is our ego—all the crap that goes in our conscious and subconscious mind. In my experience, the more I was able to reduce or even eliminate the distracting activity of my mind using tools of self-awareness, the easier it was to become fully present and connected to this original Source, which I'd come to believe was critical to my healing.

I also believe that this unseen force is constantly reminding us to stay balanced and whole, ready to help if we lose our way. And if we refuse to heed its warnings, or don't make the effort to learn how to listen for them, it finds a way to get our attention.

For years before I got sick I had heard a small voice coming from a place I could not explain, nudging me to let go of my fear and pain and to love myself and others. But I didn't know how to let go; my mind was locked in. I would put "Live my passion" on my annual goal list but could never figure out how to do that; how to stop living as some ego-driven, caterpillar-like materialist—in a context perverted by a spirit of sacrifice

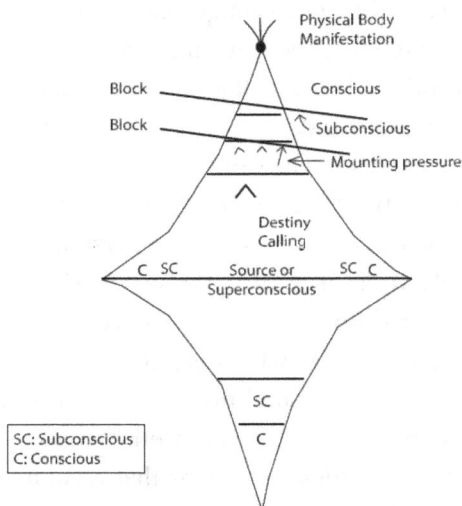

and resentment—and instead transform into a free and majestic butterfly. I knew something was coming, I knew the solution was near, but the channel through the volcano was blocked at the top. Unfortunately, or rather fortunately, that "something" was a terminal diagnosis—which to this day I believe was given to me by spirit itself.

Your Truth Will Set You Free

As I traveled the world and learned about the nature of illness, I found that in most every case, sickness emerged as a path for setting us free. I'm convinced that the immune system doesn't so much malfunction as go into hyper-mode to draw our attention to not only a dangerous situation, but to a whole new potentiality of living. This forced me to confront my disconnect with spirit, joy, love, and peace, and inspire me to confront my "monkey mind" and to start letting go of my ego. This is not your typical medical prescription but it's at the core of what I believe will be the next evolution of medicine and healing. Instead of looking at my diagnosis as dysfunctional T-cells on a rampage, I came to see them as I would rebellious workers with a legitimate grievance: Pay attention and listen or bad things will happen. Get on a new path or this journey will end.

Drawing a line between physical disease and a particular mental pattern has been difficult to demonstrate scientifically, of course, but various spiritual teachers and authors have certainly made such claims. Louise Hay's still popular 1984 book, *You Can Heal Your Life*, was inspired in large part by her experience of healing herself from cancer primarily by changing her thought patterns and confronting her childhood traumas.[10] In his 1946 book, *Autobiography of a Yogi*, the spiritual teacher Paramahansa Yogananda recounted a quote from his own teacher, Sri Yukteswar, who stated, "Thought is a force, even as electricity or gravitation. The human mind is a spark of the almighty consciousness of God. . . . whatever your powerful mind believes very intensely would instantly come to pass."[11]

More recently, German physician/oncologist Ryke Geerd Hamer[12] developed what he calls German New Medicine, which does draw a direct line between our thoughts, the brain, and specific organs. Hamer believes that diseases have a biological meaning and are meant for the benefit of the patient. Drawing from the experience of his own personal tragedy as well as two decades of scientific research, he has identified a phenomenon he calls "biological conflict-shock" that starts in our thoughts, gets hardwired into the brain, and then manifests in the body as a cancerous or necrotic condition (cell injury). Resolve the root conflict, he says, and healing is sure to follow.

Not surprisingly, the powers-that-be in various Swiss and German medical bodies rejected Hamer's approach as ineffective and even dangerous, and revoked his license. This is often the case with new and unusual treatments: very few get through the medical establishment's rigorous and conservative evaluation process. But as we have seen, the establishment has its own biases and belief systems. It wants to protect us, but it also wants to call all the shots and enforce its mechanistic, technology- and drug-driven perspective on how to deal with serious illness. From what I've seen, though, I think Hamer may be on to something.

But we're still left with the mystery of why cancer happens in some people and not others, and why cellular proliferation—a natural process that resolves itself the vast majority of the time—goes haywire. What role is played by the mind, the body, and the spirit and their relationships within that holy trinity? Where do our beliefs, thoughts, and emotions come in, whether they are conscious or subconscious? One doesn't necessarily need to experience an extreme shock for a cancerous response to be triggered. There are gradations and durations of stress that are also part of the puzzle.[13] The behavior of our immune system is either directly or indirectly controlled by the body's autonomic nervous systems, both the sympathetic ("fight and flight") and parasympathetic ("rest and rebalance"). These systems in turn are influenced by how we live and think. Our thoughts and emotions, for example, trigger chemical messengers that then modify cellular behavior. The impact of too much stress or too many negative thoughts will ripple through the body and start playing havoc with all those systems and relationships. A stressed immune system will release high levels of inflammatory cytokines, which have been linked to depression, Alzheimer's, and cancer, as well as other conditions.[13]

The scientific and medical communities largely dismiss any connection between stress reduction and cancer progression. The American Cancer Society has declared "there's no good evidence to support the idea that [psychosocial] interventions can reduce the risk of cancer, keep cancer from coming back, or help the person with cancer live longer."[14] But another group of researchers in the relatively new field of psychoneuroimmunology (PNI) is looking hard for the connection. They've been studying the relationship between the brain, our immune systems, and disease. While there's still a lot to be learned, the research is showing that our thoughts, moods, and beliefs trigger neurochemicals that cause the release of hormones which then interact with our disease-fighting cells, either helping or hindering their work. A good example of this is a paper published in 2009 reporting on a study in which the placebo response provided an effective substitute in treating psoriasis patients when compared to a traditional pharmaceutical regimen. There were three groups: one received only a fraction of their usual dose of a widely used steroid medication but was told it was full strength; the second group got the full dose some of the time and a placebo half to three-quarters of the time; the third group got a normal dose. The second and third groups had similar results; the "treatment" worked just as well on the group that "thought" they were getting a full dose as the one that really did.[15]

And so, the question for me, of course, is this: If the power of our mind works for psoriasis and other conditions, can it actually influence tumor growth in cancer? When I discovered a special issue of *Brain, Behavior, and Immunity* (March 2013) titled "Psychoneuroimmunology and cancer: A decade of discovery, paradigm shifts, and methodological innovations,"[16] I was struck by how deep the research is going. BBI has become the leading scholarly journal linking PNI and cancer. The science has evolved to the point where researchers are now looking for one common psychological pathway or signature that links to tumor development and growth—as well as its disappearance. The Institute of Noetic Sciences' *Spontaneous Remission Annotated Bibliography* has more than a thousand reported cases of cancer remission from more than 800 professional journals in 20 languages.[17] In her new book, *Radical Remission: Surviving Cancer Against All Odds,* Dr. Kelly Turner identifies nine possible factors that have been associated with the spontaneous remission of cancer, from our thoughts and emotions to nutrition and our relationships.[18] It's a brave new world of possibilities!

The brilliant addiction specialist, Dr. Gabor Maté, contends that the roots of most chronic illness—including cancer—lie not in our genes or external agents such as bacteria and viruses, but in the emotional coping patterns we learned as children and the influences of the environment we grew up in. Those patterns and influences get lodged in our cells and their impact escalates as we become adults. Our genes may create predispositions to certain diseases, but they need the right (or wrong!) conditions in which to flourish. A nurturing environment, for example, reduces the chances that these genetic inclinations will be activated. In his book *When the Body Says No: Exploring the Stress-Disease Connection,* Maté addresses the issue of "cancer personalities" and explored how certain emotions increase the risk of getting certain kinds of cancer.[19] He concluded that "While we cannot say that any personality types causes cancer, certain personality features definitely increase the risk because they are more likely to generate physiological stress [including] repression, the inability to say no, and a lack of awareness of one's own anger... repeated and multiplied over the years, they have the potential of harming homeostasis and the immune system."

In her book *The Type-C Connection: The Mind-Body Link to Cancer and your Health,* Lydia Temoshok, director of the Behavioral Medicine Program at the University of Maryland Medical School, found links between certain personality traits—including the tendency to put others' needs ahead of one's own—and cases of melanoma cancer.[20] She found a similar relationship between "Type C coping styles" and the progression of HIV.[21] It's kind of the flip side of how some aspects of Type A behavior—driven, angry, competitive—can contribute to the development of heart disease.[22]

Temoshok and others are quick to caution that they aren't necessarily proposing a direct cause and effect; they believe the processes involved are far more complicated, with genetics and environment also part of the equation. But the body-mind connection is real and we're only beginning to understand how it works. When Lydia writes about putting the needs of others before our own, and Gabor reveals the destructive power of repressed emotions, I go back to the model of ego vs. spirit. The ego—those thoughts and feelings that make up our self-identity—convinces us that we are limited to be a certain way because of all the crap that happened in our past. Life is stressful and we learned a bunch of ways to adapt and cope. But in that process, we often ended up suppressing our true needs, our natural being. Spirit doesn't play that game; it comes from a different place and knows we have a different purpose. The challenge is how to make that connection, how to open up that channel.

A big part of getting there involves self-knowledge and paying attention—listening. And a great place to start is in our own bodies. Our immune system is a beautifully designed feedback loop that generates both small warning signs and serious ultimatums—which unfortunately come after many warning signs. In my case, the warning signs were modest but persistent: being a little more tired, colds that lasted longer and were more frequent, stomach pains, diarrhea, exhaustion, trouble sleeping, infections, inflammation, digestive issues, allergic reactions, even full-body itching. Many doctors and specialists talk about the progressive occurrence of symptoms until a serious illness declares itself.[23] Like the mini-tremors Gilligan described that precede a big earthquake, they give us an indication of what is to come if we don't release the pressure. And I paid heed to none of them until I was finally given something I could not ignore: an emergency wake-up call from spirit. And, of course, it's not about just listening to your body. A lot of messages come from your heart, which spirit also uses, which often can only be heard in silence. This was a lesson I had to keep learning, over and over again.

My Truth or Your Truth?

What emerged from my deeply motivated and relentless search to discover spiritual meaning was radically different from what I'd experienced before. As I opened myself to all possibilities, and resolved to find the limits of what existed, I found something I could never have imagined, a way of being that transcended all limitations. In this realm there exists undeniable, in some cases indefinable, phenomena that are very real, that we have yet to fully integrate into our accepted worldview. Many of the encounters I had during this journey changed the very foundation of what I thought was true about the world, leaving me with an unshakable and experientially undeniable belief in a higher power and its influence in our lives.

There has always been a war within science between the old guard and the new. In genetics, for example, it has taken decades to evolve from the view that genes are fixed to one where they are dynamic and responsive, as embraced by epigenetics. Classical physics gave us the science of things that could be seen, felt, and measured, which remained unquestioned for centuries until quantum physics upended everything by claiming that those rules didn't necessarily apply in the subatomic world.

One of the most mysterious effects of quantum physics' new "science of the unseen" is the "observer effect," where the very act of observation changes whatever phenomenon is being observed. This principle is shown most clearly in what has become one of the most famous experiments in the history of physics—the double-slit experiment—where photons passing through two slits change from waves to particles (as evidenced by changes in their interference patterns) when being observed. A recent study by Dr. Dean Radin and colleagues at the Institute of Noetic Sciences, published in the mainstream journal Physics Essays (no minor feat that it was even recognized there), confirmed this connection, which is also called the "attention-modulated mind-matter interaction effect."[24] In short, it matters what we pay attention to—literally! It's a real mind blower. Science and religion are beginning to converge on the idea that the universe is not a community of separate parts but a single, indivisible whole within which we are all connected in still quite mysterious ways. One of the metaphors used to describe this enmeshed reality is that we are all entangled. (See Radin's *Entangled Minds*[25] and *The Conscious Universe*[26] and Lynne McTaggart's *The Field*.)

Some people have spun these ideas into new age memes and models (like in the book The Secret) which basically say you can create the world you want using only your mind.[26] I have yet to understand how true this is, but the mind is a very powerful factor in how we personally experience everything that happens both around us *and* inside us. What's cool about spirituality, especially for the skeptic, is that if you're willing to suspend your beliefs, go out of your box, and apply a lot—or even a little—faith, you really can experience magic. Paranormal worlds open up that your conditioned mind would never allow in, that blow up your perspective. I had a taste of this when I was younger, exploring those esoteric worlds I found on the Internet, and more of it started happening as I got deeper into my healing practice. They gave me faith that other realities were possible, and ultimately turned belief into actuality. This was critical to my healing. I was facing the impossible, and these kinds of experiences were showing me that the impossible was possible!

I started to meditate up to four hours a day shortly after my diagnosis. Meditating for this many hours changes things; actually, it changes everything. It completely transformed who I was and how my world unfolded, sometimes in unnerving ways. A profound sense of presence evolved first, just really being in my body with a spacious, alert mind— this had never happened for me. This was followed by prolonged experiences of non-attachment; walking down the street, for example, my mind would be completely clear, empty of thoughts or thinking, calm and at peace. Nothing seemed to faze me. As these states got stronger and happened longer, a space opened up and words, emotions, or visions would suddenly appear, insights into myself or unexpected connections to the world around me, such as the "free like an eagle" experience that happened at trance camp. I'd meet strangers and know their names, dream of things that then came true. People I hadn't talked to in forever would pop into my mind, the phone would ring, and there they were. Normally I'd attribute such things to chance or coincidence, but there were just too many of them.

In another such example, I was driving down Robson Street in Vancouver. It was a sunny day; I was relaxed, zoned out, when a large *thump* made me jump in my seat. A bird had hit the windshield. The impact rocked the car all the way to the steering wheel, and I had that emotional, sinking feeling of killing another creature. It seemed clear as day, but at the same time there was something oddly vague about it. Pulled out of my "trance," I immediately realized that while the experience felt completely real, it hadn't actually happened: no bird, no thump, no mark on the window. My heart rate shot up, though I could still see the image in my mind: the bird flying into the top right corner of the window, a purple-green flash of its feathers as the bird rolled over the window and off the side of the car. But it didn't happen.

More alert now, I kept driving down the street, made a few turns, and then a bird really did fly into the window. THUMP! It hit hard, followed by a purple-green flash as it rolled across the windshield and off the side of the car, exactly as I had "seen" it earlier. I pulled over immediately. My heart was pounding. I stared out the window, unwilling to accept this impossible event. Was all the stress I'd been feeling starting to get to me and now I was seeing things? Was there *really* a bird out there? I wanted to drive away and pretend it never happened, but somewhere inside I knew that this was a sign, that it had happened for a reason. I'd been yearning for contact with something bigger than me, but I was also afraid of it—it was something I couldn't control, and didn't know how to let in.

I was about to pull away when a voice came up from deep inside: "You must look." I had promised myself to do whatever it took to figure out this impossible mission of curing my terminal illness, to go however far down the rabbit hole I needed to go, and this experience was down there with the rabbit. I threw the car back into park, got out, and walked around to the back of the car in what seemed like slow motion. When I finally reached the back, I saw it: a dead pigeon. I kneeled down and reached out to touch it. It was still warm. "I'm sorry," I said. I stopped meditating that day, and wouldn't return to the cushion with the same intensity until four years later. I felt like I'd reached a place that was very special, that I'd achieved what I'd set out to do, and also that I'd started to spend too much time in an altered state. It was time to focus more on other matters.

To Be or Not to Be, or...

One of the radically transformative ideas that Gilligan presented at Trance Camp was that of duality, the essence of how our physical and perceptual world is structured (light/dark, hot/cold, open/closed, here/there, in view/ out of view, etc.), is a construct of the mind. The principle of *entanglement* (also referred to as "nonlocality"), a core concept in quantum physics,[27] essentially states that once two things have been connected (like a pair of photons), they never lose that connection even if they exist separately at a great distance from each other. Tweak one and the other also "feels" it. Said another way, entanglement is an experimentally verified and accepted property of nature that says everything that has ever come into contact is forever connected. And since the current model of the universe says it all started from a single point that exploded with the Big Bang, making us all part of the same stardust, well, you can see where this could go. It opens the door to seeing the world in a profoundly different way where duality doesn't exist. It is only our mind that differentiates and delineates things, as it has been trained to do from birth. And so, the isolation and separation we experience is actually an illusion.

"Pick a spot where you have enough room to step forward and backward and side-to-side without bumping into everyone," Gilligan instructs us in a calm, entrancing voice.

"Now pick a statement ..."

Where others picked "I love myself" or "I am free from my trauma," I chose something controversial—*I am god*—as a way of testing how I felt about this.

Gilligan defined four positions around an imaginary center point on which we stood, then led us into a trance where we reached a place of openness and receptivity where we could truly experience whatever was about to happen. We then stepped forward in front of our circles where our statements were true.

"I am god."

Wow! This felt like a powerful place. From here the world looked different. Fear vaporized.

We stepped back to center, then to the back, adding a "not" to our statement to make it untrue.

"I am not god."

Okay, that sounds about right. I am flesh and blood and comfortable with being just human.

Back to the center, and now to the right: "I am neither god nor not god."

That was more complicated, and after scanning for a reference point,
I settled on emptiness. I had this very cool sensation of everything disappearing.

We stepped back to center again and finally to the left.

"I am both god and not god."

My head jammed up. I couldn't think of how that worked and my mind went blank.

Then back to the center, where Gilligan now explained that all the statements we had made were true. And BAM! It hit me. It opened the door for a new way of being, where duality simply slipped away, replaced by infinite possibility. It's really hard to describe; one of those things that you have to experience to understand. But it was a turning point for me; life was never the same after that.

The intention of the exercise was to help us experience a state of being beyond the limitations and judgments of our minds, a way of seeing between the lines, of realizing the power we had to embrace or limit possibilities using rules we'd placed on a world that had very few of them. For me, *this* was spirituality, this was the portal to the magic of the universe. Until this point I'd been stuck in a dilemma between the idea of things being fixed in place and one that said I could change anything though belief. This was the experience I needed to set me free. It was just a matter of my point of view and the reality I wanted for myself. I now knew I had a choice.

Changing the Soil

In the early days of my illness, I instinctively knew that the way out had to start from under the surface. It wasn't until I'd done some research on what might be going on there that I finally decided to act on this knowledge.

My heart was calling for things that I didn't allow, emotions like self-love that I had blocked; creative expressions I had suppressed; deep relationships and community with others I had resisted; a spirit of playfulness and belief in my potential that I barely acknowledged. I was like a seed from a massive maple tree that had fallen at the wrong time of year or into the wrong type of soil. The potential was there but the conditions to flourish were not. My mind had blocked them and now I was paying the price. The road ahead was starkly clear: change or die. And so, I resolved to re-till the soil, to change everything about the person I had become. I embraced the deepest calling I had ever felt to connect with spirit and started doing whatever it took to grow as a person and manifest my dreams. I began by eliminating the frictions that had made me sick, using spirit as my guide:

(1) Go under the surface and acknowledge the parts of myself that had created my illness, and also the parts that wanted love, joy, and peace as a way of being.

(2) Accept spirit as real and surrender to it, asking for its wisdom and its presence.

(3) Create openings for spirit to show up.

(4) Fearlessly move toward the things that spirit challenges me to do.

(5) Be real! Acknowledge when I deviate from its plan and adjust.

In short, I dedicated my life to the service of spirit, and it was never boring. It was a process that taught me something new every day. I used it as a foundation for everything else in my healing program, and found a whole new way of living.

Endnotes

1 Gilligan, Stephen G. "Symptom phenomena as trance phenomena." *Developing Ericksonian therapy: State of the art* (1988): 327–352.

2 "An Interview with Dr. Stephen Gilligan (2004)." Interview by Chris Collingwood. *Ericksonian Hypnosis Training.* Inspiritive Pty Ltd, 2004. Web. 8 Oct. 2014. http://www.ericksonianhypnosis.com.au/steve_gilligan_interview_2004.htm

3 Erickson, Milton H., and Ernest L. Rossi. "Autohypnotic Experiences of of Milton H. Erickson." *American Journal of Clinical Hypnosis* 20.1 (1977): 36–54.

4 Barnes, Jo, et al. "Hypnotherapy for smoking cessation." *Cochrane Database Syst Rev* 10 (2010). Web. 8 Oct. 2014.

5 Spiegel, Sharon B. "Current Issues in the Treatment of Specific Phobia: Recommendations for Innovative Applications of Hypnosis." *American Journal of Clinical Hypnosis* 56.4 (2014): 389–404. Web. 8 Oct. 2014.

6 Dilts, Robert, Tim Hallbom, and Suzi Smith. *Beliefs: Pathways to Health & Well-being.* Portland, Or.: Metamorphous, 1990. Print.

7 Chong, Dennis K., and Jennifer K. Smith Chong. "The Ontology of Malignancy and the Possibility to Turn it." (n.d.): Paper to the 1997 NLP Conference. Toronto: NLP Conference, Valhalla Inn, 1997. *Neurosemantic Programming*. Web. 8 Oct. 2014.

8 Targ, Elizabeth F., and Ellen G. Levine. "The Efficacy of a Mind-body-spirit Group for Women with Breast Cancer: A Randomized Controlled Trial." *General Hospital Psychiatry* 24.4 (2002): 238-48. Web. 11 Oct. 2014.

9 Smith, Huston. *The Religions of Man*. 1st ed. New York, NY: Harper and Bross, 1958. Print.

10 Hay, Louise L. *You Can Heal Your Life*. Santa Monica, CA: Hay House, 1984.

11 Yogananda, Paramahansa. *Autobiography of a Yogi*. New York: Philosophical Library, 1946. Print.

12 Ventegodt, Søren, Niels Jørgen Andersen, and Joav Merrick. "Rationality and Irrationality in Ryke Geerd Hamer's System for Holistic Treatment of Metastatic Cancer." *The Scientific World Journal* 5 (2005): 93–102. Web. 11 Oct. 2014.

13 Marshall, Gailen D. "The Adverse Effects of Psychological Stress on Immunoregulatory Balance: Applications to Human Inflammatory Diseases." *Immunology and Allergy Clinics of North America* 31.1 (2011): 133–40. Web. 11 Oct. 2014.

14 "Attitudes and Cancer." *Attitudes and Cancer*. American Cancer Society, 31 Mar. 2014. Web.09 Oct. 2014.

<http://www.cancer.org/treatment/treatmentsandsideeffects/emotionalsideeffects/attitudes-and-cancer>.

15 Ader, R., M. G. Mercurio, J. Walton, D. James, M. Davis, V. Ojha, A. B. Kimball, and D. Fiorentino. "Conditioned Pharmacotherapeutic Effects: A Preliminary Study." *Psychosomatic Medicine* 72.2 (2010): 192–97. Web. 5 Nov. 2014.

16 Mcdonald, Paige Green, Mary O'Connell, and Susan K. Lutgendorf. "Psychoneuroimmunology and Cancer: A Decade of Discovery, Paradigm Shifts, and Methodological Innovations." *Brain, Behavior, and Immunity* 30 (2013): S1-S9. Web. 18 Oct. 2014.

17 O'Regan, Brendan, and Caryle Hirshberg. *Spontaneous remission: an annotated bibliography*. Sausalito: Institute of Noetic Sciences, 1993. Print.

18 Turner, Kelly A. *Radical Remission: Surviving Cancer against All Odds*. New York, NY: HarperOne, 2014. Print.

19 Maté, Gabor. *When the Body Says No: Exploring the Stress-disease Connection*. Hoboken, NJ: J. Wiley, 2011. Print.

20 Temoshok, Lydia, and Henry Dreher. *The Type C Connection: The Mind-body Link to Cancer and Your Health*. New York: Plume, 1992. Print.

21 Temoshok, Lydia R., Shari R. Waldstein, Rebecca L. Wald, Alfredo Garzino-Demo, Stephen J. Synowski, Lingling Sun, and James A. Wiley. "Type C Coping, Alexithymia, and Heart Rate Reactivity Are Associated Independently and Differentially with Specific Immune Mechanisms Linked to HIV Progression." *Brain, Behavior, and Immunity* 22.5 (2008): 781-92. Web. 18 Oct. 2014.

22 Trigo, Miguel, Danilo Silva, and Evangelista Rocha. "Factores Psicossociais de Risco na Doença Coronária: Para Além do Comportamento Tipo A [19]." *Rev port cardiol* 24.2 (2005): 261–281. Web. 18 Oct. 2014.

23 Confino-Cohen, Ronit, Gabriel Chodick, Varda Shalev, Moshe Leshno, Oded Kimhi, and Arnon Goldberg. "Chronic Urticaria and Autoimmunity: Associations Found in a Large Population Study." *Journal of Allergy and Clinical Immunology* 129.5 (2012): 1307–313. Web. 18 Oct. 2014.

24 Radin, Dean, et al. "Psychophysical interactions with a double-slit interference pattern." *Physics Essays* 26.4 (2013): 553-566.

25 Radin, Dean. *Entangled minds: Extrasensory Experiences in a Quantum Reality.* Simon and Schuster, 2009.

26 27.Byrne, Rhonda. *The Secret.* New York: Atria, 2006.

27 Horodecki, Ryszard, Michał Horodecki, and Karol Horodecki. "Quantum Entanglement." *Reviews of Modern Physics* 81.2 (2009): 865–942. Web. 21 Oct. 2014.

Brains and Beyond:
The Unfolding Vision of Health and Healing*

LARRY DOSSEY

*The notion of a separate organism is clearly an abstraction,
as is also its boundary. Underlying all this is unbroken wholeness
even though our civilization has developed in such a way
as to strongly emphasize the separation into parts.[1]*
— DAVID BOHM AND BASIL J. HILEY,
The Undivided Universe

"I suddenly developed a severe headache in the back of my head," the nurse said tearfully. "It was so painful I could not function and had to leave work. This was strange, because I never have headaches. When I reached home and was lying in bed, the phone rang. I learned that my beloved brother had been killed from a gunshot wound to the back of his head, the same place my terrible headache was located. My headache began at the same time the shooting occurred."

The woman was a prominent nurse leader at a major hospital in northern California. The occasion was a Q & A session following an address I had given to senior staff of the hospital consortium to which her hospital belonged. My topic was the importance of empathy, compassion, and caring in healing and healthcare. I had reviewed empirical evidence suggesting that empathy and compassion are more than vaporous emotions that float in our bodies somewhere above our clavicles. They are part of our biological makeup, I suggested. While empathy and compassion arise when we are in the presence of another person, as when a nurse or physician is at the bedside of a patient, evidence suggests their effects may also be felt between individuals at a distance, beyond the reach of the senses. Distant individuals often share feelings, sensations, and thoughts, particularly if they are emotionally close. These experiences, I explained,

* (This chapter is based on an address by Larry Dossey, MD, to the "Behind and Beyond the Brain" conference, the 11th Symposium of the Bial Foundation, in Porto, Portugal, April 2, 2016)

are often called *telesomatic* events. Hundreds of such cases have been reported over the years, but have been largely ignored.

This discussion had prompted the nurse to reveal her experience to several hundred of her colleagues in the audience. "Now I have a name for what happened between my brother and me," she said. "Now I can talk about it." Her story riveted the audience. When she finished, she was not the only person in the room in tears.

This woman's story is, of course, "only an anecdote." "Anecdote" comes from the Greek *anekdota*, "things unpublished." Our lives are comprised of anecdotes—stories, happenings, events, and experiences that are all unpublished. Our existence does not unfold as a series of controlled, publishable scientific studies. It is when our experiences form patterns that are shared by others that we should pay attention to the possible messages they may convey.

Experiences such as these are not uncommon. They suggest a unity and connectedness between biological systems that transcend separation in space.

A growing body of evidence supports this invisible connectivity at several levels of biological complexity. This evidence goes beyond the etymology of "anecdote," for it has indeed been published in peer-reviewed journals and is now a part of the scientific record.

Distant Mental Interactions with Living Systems (DMILS)

Experiments generally known as DMILS—*d*istant *m*ental *i*nteractions with *l*iving *s*ystems—involve a wide variety of entities such as whole humans, organs, cells, microbes, plants, and animals. In these studies, individuals use their intentions to influence biological functions in humans, the growth rates of bacteria and fungi in test tubes and Petri dishes, the rate of wound healing in mice, the healing of transplanted cancers in mice, the function of cells in tissue cultures, the germination rates of seeds, the growth rates of seedlings; and many other phenomena. Two examples follow.

Gronowicz and colleagues assessed the effect of Therapeutic Touch (TT) on the proliferation of normal human cells in culture, compared to sham and no-treatment controls. This non-touch technique, which emphasizes healing intentions, was administered twice a week for 2 weeks. Compared to untreated controls, TT significantly stimulated proliferation of fibroblasts (cells that produce collagen and are important in wound healing), tenocytes (tendon cells), and osteoblasts (bone cells) in culture (P = 0.04, 0.01, and 0.01, respectively). These data were obtained by sophisticated techniques such as immunocytochemical staining for proliferating cell nuclear antigen (PCNA). The researchers concluded, "A specific pattern of TT treatment produced a significant increase in proliferation of fibroblasts, osteoblasts, and tenocytes in culture. Therefore, TT may affect normal cells by stimulating cell proliferation."[2]

In 10 controlled experiments, researcher William Bengston tested the effect of "healing with intent" on laboratory mice. (See his chapter in this book.) In 8 of these experiments, mice were injected with mammary adenocarcinoma (breast cancer) cells. In 2 experiments, mice with methylcholanthrene-induced sarcomas were used. The fatality rate for both cancers in mice, if untreated, is 100 percent. The healers were faculty and student volunteers. Although they had no previous experience or belief in healing with intent and were often skeptical of such, they were drilled extensively in the healing technique. Treatment length was from 30 to 60 minutes, delivered daily to weekly until the mice were cured or died. They were successful in producing full cures in approximately 90 percent of the mice. When mammary adenocarcinoma cells were re-injected into cured mice, the cancer would not take, suggesting that an immune response had been stimulated during treatment. The proximity of the volunteer healers to the cages of the mice varied from on site to approximately 600 miles. Thus, Bengston notes, "[T]hese effects were at times brought about from a distance that defies conventional understanding," suggesting that a nonlocal process was at work. This series of studies, conducted at several academic centers, suggests that healing through intent can be predictable, reliable, and replicable.[3,4,5,6]

However, the DMILS field is too extensive to be reviewed here. These studies are described and summarized in readily available sources; a recent review must suffice.[7,8,9,10,11,12,13,14,15,16,17,18] In a 2015 meta-analysis of this field, consciousness researcher Chris A. Roe and his colleagues at the University of Northampton examined 106 "noncontact healing studies"—57 involving whole humans and 49 involving non-whole humans (tissues, cells) and nonhumans (animals, plants, etc.). All the various healing methods employed in these experiments incorporated an intention to heal. The researchers concluded, "Results in the active condition exhibit a significant improvement in wellbeing relative to control subjects.... [Results] do not seem to be susceptible to placebo and expectancy effects.... The effect size is small, but statistically significant."[19]

To reiterate, non-humans such as cells, plants, microbes and biochemical reactions presumably do not think positively or symbolically and are therefore not subject to suggestion and expectation. If in controlled experiments these entities respond to intentions, presumably the placebo response is not responsible, but the influence of the thoughts and intentions of the healer.

This generalization requires qualification. In humans, placebo effects are believed to be mediated by the empathy, compassion, likeability, and trustworthiness manifested by a physician. Thus, veterinarian and placebo researcher F. D. McMillan states, "To the extent that animals form such perceptions...it is reasonable to posit a similar influence of placebo effects in animal health care."[20] There is evidence that certain non-human animals can manifest placebo effects through operant conditioning. For example, Ader and Cohen paired an immunosuppressive drug (cyclophosphamide) with a neutral stimulus (a saccharine solution) in mice with a lupus-like disease. When only the neutral stimulus was later given, the result was immunosuppression, suggestive of a placebo response.[21,22] Moreover, a body of research demonstrates healthy effects in animals from visual and tactile contact from a human, involving rabbits, dogs, horses, dairy cows, and sows.

How, then, can placebo responses be differentiated from our hypothesized effects of healing intentionality? The reasons are straightforward. Many of the relevant studies do not involve animals at all, but cells, tissues, plants, microbes, and chemical reactions. Moreover, intentionality effects do not depend on proximity to a subject. Many of the experiments suggesting distant healing effects have been done remotely, beyond sensory contact. This suggests that a nonlocal phenomenon is at play, as opposed to the *local*, sensory-mediated mechanisms believed to underlie placebo responses in humans and higher animals. Therefore, if animals are not involved as test subjects, and if sensory-mediated contact is by-passed, placebo effects would appear to have been eliminated.[23]

Cell-to-Cell Connections

In 2009, a team of Italian researchers led by neuroscientist Rita Pizzi repeatedly demonstrated that when one batch of human neurons was stimulated by a laser beam, a distant batch of neurons registered similar changes, although the two were completely shielded from each other. The researchers concluded, "[O]ur experimental data seem to strongly suggest that biological systems present nonlocal properties not explainable by classical models."[24]

In 2007 researcher Ashkan Farhadi and colleagues at Rush University Medical Center in Chicago examined whether cells in separate containers could communicate with each other. They exposed one container of intestinal epithelial "inducer" cells to hydrogen peroxide and assessed the damage done to them. Another batch of "detector" cells was placed in a separate container and was not exposed to hydrogen peroxide. Even though there was no obvious way the two batches of cells could communicate, the detector cells demonstrated damage similar to the inducer cells. "These findings," the researchers said, "provide evidence in support of a non-chemical, non-electrical communication."[25]

In 2013 researcher Victor B. Chaban and his colleagues at UCLA School of Medicine, demonstrated "physically disconnected non-diffusable cell-to-cell communication" between neuroblastoma cancer cells and normal neurons, when both are shielded, preventing any known means of communication.[26]

Brain-to-Brain Connections

In 1965 researchers T. D. Duane and Thomas Behrendt decided to test anecdotal reports that identical twins share feelings and physical sensations even when far apart. In two of fifteen pairs of twins tested, eye closure in one twin produced not only an immediate alpha rhythm in his own brain, but also in the brain of the other twin, even though he kept his eyes open and sat in a lighted room.[27]

The publication of this study in the prestigious journal *Science* evoked enormous interest. Ten attempted replications soon followed, by eight different research groups around the world. Of the ten studies, eight reported positive findings, published in mainstream journals such as *Nature* and *Behavioral Neuroscience*.[28,29,30,31,32,33,34,35 36,37]

In the late 1980s and 1990s, a team headed by psychophysiologist Jacobo Grinberg-Zylberbaum at the University of Mexico published experiments that, like most of the previous studies, demonstrated correlations in the EEGs of separated pairs of individuals who had no sensory contact with each other.[38,39,40] Two of the studies were published in the prominent journals *Physics Essays* and *International Journal of Neuroscience*, drawing further attention to this area.[41,42,43]

Experiments in this field became increasingly sophisticated. In 2003 Jiri Wackerman, an EEG expert from Germany's University of Freiberg, attempted to eliminate all possible weaknesses in earlier studies and applied a refined method of analysis. Following his successful experiment, he concluded, "We are facing a phenomenon which is neither easy to dismiss as a methodological failure or a technical artifact nor understood as to its nature. No biophysical mechanism is presently known that could be responsible for the observed correlations between EEGs of two separated subjects."[44]

As fMRI brain-scanning techniques matured, these began to be employed, with intriguing results. Psychologist Leanna Standish at Seattle's Bastyr University found that when one individual in one room was visually stimulated by a flickering light, there was a significant increase in brain activity in a person in a distant room.[45]

In 2004, three new independent replications were reported, all successful—from Standish's group at Bastyr University,[46] from the University of Edinburgh,[47] and from researcher Dean Radin and his team at the Institute of Noetic Sciences.[48]

Person–to–Person Connections

Evidence that our thoughts, emotions, and behaviors may influence some-
one remotely has surfaced in recent analyses of social networks. The
precise mechanism of these phenomena is currently unknown. James H.
Fowler, a political scientist at the University of California, San Diego,
and Nicholas A. Christakis, a physician and social scientist at Harvard
Medical School, published a provocative article in 2008 in the *British
Medical Journal*, titled "Dynamic Spread of Happiness in a Large Social
Network."[49] Christakis states, "[H]appiness is more contagious than previously thought.... Your happiness depends not just on your choices and
actions, but also on the choices and actions of people you don't even know
who are one, two and three degrees removed from you.... Emotions have
a collective existence — they are not just an individual phenomenon."[50]

From 1983 to 2003, Fowler and Christakis collected information
from 4,739 people enrolled in the well-known Framingham Heart Study
and from several thousand other individuals with whom they were connected—spouses, relatives, close friends, neighbors and co-workers.
They found, says Fowler, that, "[I]f your friend's friend's friend becomes
happy, that has a bigger impact on you being happy than putting an extra
$5,000 in your pocket." The idea that the emotional state of your friend's
friend's friend could profoundly affect your psyche created a sensation in
the popular media. As a *Washington Post* journalist put it, "[E]motion can
ripple through clusters of people who may not even know each other."[51]

It's not just happiness that gets around. The team also found that depression, sadness, obesity, drinking and smoking habits, ill-health, the
inclination to turn out and vote in elections, a taste for certain music or
food, a preference for online privacy, and the tendency to think about sui-
cide are also contagious.[52,53]

Christakis and Fowler published their findings about the spread of
obesity in large social networks in the influential *New England Journal
of Medicine*. They showed that obesity in people you don't know and
have never heard of could ricochet through you. They attributed the con-
tagiousness of obesity to a "social network phenomenon" without proposing any specific physiological or psychological mechanism.[54] To label
something, however, is not to explain it, and to merely call this sort of
thing a "social network phenomenon" has all the explanatory value of

saying "what happens happens." In the commentary that accompanied their *NEJM* article, the experts who weighed in took the same tack. They discussed the genetic factors that influence obesity and the connections within and between cells in an individual that may contribute to over-weight, but they too were mute about how distant humans might influence one another when they are beyond sensory contact.

Some suggest that the ripples work through the action of mirror neu-rons, which are brain cells believed to fire both when we perform an ac-tion ourselves and when we watch someone else doing it. But when peo-ple are remote from each other, there is no one to watch, and therefore no stimulus for the mirror neurons to fire. Others suggest that the spread is through mimicry, as when people unconsciously copy the facial expres-sions, body language, posture, and speech of those around them. There is a hint of desperation in these attempts to find some sneaky physical factor that mediates changes between distant individuals. But when all is said and done, Fowler and Christakis say they don't really know how happi-ness, obesity, etc. spread.[55]

The fact that your friend's friend's friend, someone you've neither seen nor heard of, is affecting your health has begun to rattle many of the gatekeepers in medicine. This field may be a bomb with a delayed fuse that is getting ready to explode in the very heart of materialistic medicine. A few medical insiders are raising the possibility that something hereto-fore unthinkable may be going on, such as a nonlocal, collective aspect of consciousness that links distant individuals. Among them is Dr. Robert S. Bobrow, a courageous clinical associate professor in the Department of Family Medicine at New York's Stony Brook University. In dis-cussing the spread of obesity in his article "Evidence for a Communal Consciousness" in *Explore* in 2011, he says, "Frankly, obesity that de-velops from social connection, without face-to-face interaction, suggests emotional telepathy."[56]

If these experiments don't take your breath away, they should. They suggest that human isolation is a myth, and that human consciousness can manifest in the world beyond the brain. We are linked, united, entangled.

Do these person-to-person connections represent genuinely nonlocal phenomena? Are they on the same order as the cell-to-cell events demon-strated in the experiments of Pizzi, Farhadi, and Chaban? Currently no one knows for certain, as mentioned, and further research will hopefully clarify these important questions. On balance, however, as Bohm and Hiley state in the epigraph, "The notion of a separate organism is clearly an abstraction, as is also its boundary."

Telesomatic Events

But if you stop clinging to coincidence
and try explaining this trumpery affair,
you might shatter one kind of world.[57]
— J. B. PRIESTLEY, *Man & Time*

Almost forgotten amid this flurry of research are hundreds of case reports such as the experience of the nurse above, which suggest a person-to-person form of communication that appears genuinely nonlocal. In them, individuals experience similar sensations or actual physical changes, even though they may be separated by great distances. Berthold E. Schwarz, an American neuropsychiatrist, documented many of these instances. In the 1960s he coined the term *telesomatic* to describe these events, from Greek words meaning "distant body."[58] The term is apt, because these events suggest that a shared mind is bridging two bodies. Most cases go unreported, however, because there is no accepted explanatory mechanism for them, and because of the social stigma that can result from discussing them publicly.

These happenings have an interesting pedigree. A typical example was described by the English social critic John Ruskin (1819–1900). It involved Arthur Severn, a famous landscape painter who was married to Ruskin's cousin Joan. Severn awoke early one morning and went to a nearby lake for a sail, while Joan remained in bed. She was suddenly awakened by the sensation of a severe, painful blow to the mouth, of no apparent cause. Shortly thereafter her husband Arthur returned, holding a cloth to his bleeding mouth. He reported that the wind had freshened abruptly and caused the boom to hit him in the mouth, almost knocking him from the boat, at the estimated time his wife felt the blow.[59]

A similar instance was reported in 2002 by mathematician-statistician Douglas Stokes. When he was teaching at the University of Michigan, one of his students reported that his father was knocked off a bench one day by an "invisible blow to the jaw." Five minutes later his dad received a call from a local gymnasium where his wife was exercising, informing him that she had broken her jaw on a piece of fitness equipment.

David Lorimer, a shrewd analyst of consciousness and a leader of the Scientific and Medical Network, an international organization based in the U.K., has collected many telesomatic cases in his wise book *Whole in One*.[60] Lorimer is struck by the fact that these events occur mainly between people who are emotionally close. He makes a strong case for what he calls "empathic resonance," which he believes links individuals across space and time.

The late psychiatrist Ian Stevenson (1918–2007), of the University of Virginia, investigated scores of instances in which distant individuals experience similar physical symptoms. Most involve parents and children, spouses, siblings, twins, lovers, and very close friends.[61] Again, the common thread is the emotional closeness and empathy experienced by the separated persons.

In a typical example reported by Stevenson, a mother was writing a letter to her daughter, who had recently gone away to college. For no obvious reason her right hand began to burn so severely she had to put down her pen. She received a phone call less than an hour later informing her that her daughter's right hand had been severely burned by acid in a laboratory accident at the same time that she, the mother, had felt the burning pain.[62]

In a case reported by researcher Louisa E. Rhine, a woman suddenly doubled over, clutching her chest in severe pain, saying "Something has happened to Nell, she has been hurt." Two hours later the sheriff arrived to inform her that her daughter Nell had been involved in an auto accident, and that a piece of the steering wheel had penetrated her chest.[63]

Twin Connections

Guy Lyon Playfair, a consciousness researcher in Great Britain, is the author of the important book *Twin Telepathy*.[64] He has collected a variety of documented telesomatic cases involving twins and non-twin siblings.

One case involved the identical twins Ross and Norris McWhirter, who were well known in Britain as co-editors of the *Guinness Book of Records*. On November 27, 1975, Ross was fatally shot in the head and chest by two gunmen on the doorstep of his north London home. According to an individual who was with his twin brother Norris, Norris reacted in a dramatic way at the time of the shooting, almost as if he had been shot by an invisible bullet.[65]

Skeptics invariably dismiss cases such as these as coincidence, but many are hard to squeeze into this category. An example reported by Playfair concerns four-year-old identical twins Silvia and Marta Landa, who lived in the village of Murillo de Río Leza in northern Spain. The Landa twins became celebrities in 1976 after being featured in the local newspaper following a bizarre event. Marta had burned her hand on a hot clothes iron. As a large red blister was forming, an identical one developed on the hand of Silvia, who was away visiting her grandparents at the time. Silvia was taken to the doctor, unaware of what had happened to her sister Marta. When the two little girls were united, their parents saw that the blisters were the same size and on the same part of the hand.

It wasn't the first time this sort of thing had happened. If one twin had an accident, the other twin seemed to know about it, even though they were nowhere near each other. Once, when they arrived home in their car, Marta hopped out and ran inside the house, but suddenly complained that she could not move her foot. While this was happening, Silvia had got tangled up with the seat belt and her foot was stuck in it. On another occasion when one of them had misbehaved and was given a smack, the other one, out of sight, immediately burst into tears.

Members of the Madrid office of the Spanish Parapsychological Society got wind of the burned-hand incident, and decided to investigate. Their team of nine psychologists, psychiatrists, and physicians descended on the Landa house, with the full cooperation and approval of the twins' parents. They had hardly arrived when a typical trade-off incident happened to the little twins. When Marta accidentally banged her head on something, it was her sister Silvia who began to cry. The researchers got to work with a series of tests disguised as fun games for the twins. This meant the little girls had no idea they were involved in an experiment.

While Marta stayed on the ground floor with her mother and some of the researchers, Silvia went with her father and the rest of the team to the second floor. Everything that happened on both floors was filmed and tape-recorded. One of the psychologists played a game with Marta, using a glove puppet. Silvia was given an identical puppet, but no game was played. Downstairs, Marta grabbed the puppet and threw it at the investigator. Upstairs, at the same time, Silvia did the same.

One of the team's physicians next shined a bright light into Marta's left eye, as part of a simple physical check-up. When she did this four times, Silvia began to blink rapidly as if trying to avoid a bright light. Then the doctor did a knee-jerk reflex test by tapping her left knee tendon three times. At the same time, Silvia began to jerk her leg so dramatically that her father, unaware the test was going on downstairs on Marta, had to hold it still. Then Marta was given some very aromatic perfume to smell. As she did so, Silvia shook her head and put her hand over her nose. Next, still in different rooms, the twins were given seven colored discs and were asked to arrange them in any order they liked. They arranged them in exactly the same order.

There were other tests as well. The team rated all but one of them as "highly positive" or "positive."

The Landa tests confirmed what many researchers have found—that children are more prone than adults to this sort of thing, and that results are more likely to be positive when experiments are done not in sterile, impersonal labs, but in the natural habitat of the subjects and in a relaxed, supportive environment. This latter lesson has often been flagrantly ignored in consciousness research by experimenters who should know better. Researchers have had to learn repeatedly the importance of *ecological validity*—the principle that what is being tested should be allowed to unfold as it does in real life.

Although telesomatic exchanges are by no means limited to twins, they are frequent among them. As Playfair states, in twins we see "the telepathic signal at full volume, as it were, at which not only information is transmitted at a distance but so are emotions, physical sensations and even symptoms such as burns and bruises."[66] Even so, he has found that only around 30 percent of identical twins have these experiences, but in those who do the phenomena can be mind-boggling.[67] Emotional closeness is an essential factor in the twin connection. Also, having an extraverted, outgoing personality has been shown to facilitate the link. And, as we see in the above examples, what twins seem to communicate best is bad news—depression, illness, accidents or death.

Era III Medicine:
The Next Step for the Mind-Body Field

We can take a socio-historical approach in sorting out the panoply of therapies currently available in the health professions.[68] Let's begin this perspective with the advent of modern, scientific medicine, which medical historians date to around the decade of the 1860s. About this time medicine began gradually to take on the complexion we see today. We can designate this as Era I medicine or physical medicine, because of its overwhelming reliance on physical measures such as drugs and surgical procedures, which continues to this day. In Era I, the mind is assumed to play a nonexistent or negligible role in health and illness.

Shortly after World War II, Era II medicine or mind-body medicine began to unfold. This was a radical departure from Era I, because in Era II the various expression of consciousness, such as thought and emotions, were acknowledged as causal factors in health within single individuals. These factors were not trivial; they might sometimes make the difference in life and death. The mind-body perspective did not negate or displace the physical focus of Era I, however, but overlapped with the drugs-and-surgery emphasis.

We are now seeing the birth of Era III medicine, the next great step in healing. Era III medicine acknowledges the intrapersonal effects of thoughts and emotions of Era II, but recognizes interpersonal effects as well. In other words, in Era III medicine one's thoughts, emotions, beliefs, and intentions can affect not just one's own body, but other individuals as well.

The premise underlying Era III is that minds at some level are connected and unitary. I've called Era III *nonlocal* medicine, leaning on the concept of nonlocality in modern physics. According to experimental evidence that is practically unchallenged, distant particles that were originally in contact behave as if they are a single particle, even though they may be widely separated at arbitrary distances.[69] When one changes they both change, instantly and to the same degree.[70]

That's not to say that the nonlocality of physical particles such as electrons or photons can account for the remote connectedness of minds, or that mental phenomena can be reduced to the behavior of subatomic particles, but that both particles and people display a kind of connectedness that defies separation in space and time. "Nonlocal" is a fitting description not only for particles but for minds as well, because "nonlocal" literally means "not in a place." Yet we should not equate the two phenomena; we may be dealing with accidental correlations of terminology—analogies, not homologies. Further scientific investigation may clarify this important issue.

Nonlocal Mind and Health

Nonlocal expressions of consciousness are frequently concerned with survival and therefore health. When information is shared between humans remotely, it is commonly about health risks, such as impending physical dangers, as we've seen. The quintessential example is a mother who "just knows" her child is in danger and takes measures to prevent harm, as in the following example from the archives of the Rhine Research Center in Durham, North Carolina.

Amanda, a young mother living in Washington State, awoke one night at 2:30 A.M. from a nightmare. She dreamed that a large chandelier that hung above their baby's bed in the next room fell into the crib and crushed the infant. In the dream, as she and her husband stood amid the wreckage, she saw that a clock on the baby's dresser read 4:35 A.M. The weather in the dream was violent; rain hammered the window and the wind was blowing a gale. The dream was so terrifying she roused her husband and told him about it. He laughed, told her the dream was silly, and urged her to go back to sleep, which he promptly did. But the dream was so frightening that Amanda went to the baby's room and brought the child back to bed with her. She noted that the weather was calm, not stormy as in the dream.

Amanda felt foolish—until around two hours later, when she and her husband were awakened by a loud crash. They dashed into the nursery and found the crib demolished by the chandelier, which had fallen directly into it. Amanda noted that the clock on the dresser read 4:35 A.M.

and that the weather had changed. Now there was howling wind and rain. This time, her husband was not laughing.

Amanda's dream was a snapshot of the future—down to the specific event, the precise time it would happen, and a change in the weather.[71]

The image of consciousness flowing from this and thousands of similar cases is a *nonlocal* one, in which some aspect of consciousness appears unconfined to specific points in space, such as brains and bodies, or time, such as the present.

Unlike Amanda's experience, however, the information we gain nonlocally is often unconscious. The information may be nonlocal with respect not only to space, but to time as well, as mentioned. For example, an individual may cancel a travel reservation because of a vague gut feeling that something is not right, or that something ominous is going to happen, not because he actually foresees a specific event. This may be one reason why occupancy rates are statistically lower on the day of train wrecks compared to non-accident days.[72] Nonlocal awareness of dire future events may also account for why the overall vacancy rate on the four doomed planes on September 11 was nearly 80 percent.

From a survival perspective, it may be an advantage for information that is nonlocally acquired to be unconscious. Thinking, analyzing, and reasoning take time. In emergencies, instant reflexive action can save a life.

If minds are nonlocal in space and time, they are unbounded. This implies that at some level they come together with other minds and form a collective or universal mind. Nobel physicist Erwin Schrödinger, whose wave equation lies at the heart of quantum physics, was interested in this possibility and believed it to be true. As he put it, "To divide or multiply consciousness is something meaningless.[73] There is obviously only one alternative, namely the unification of minds or consciousness.... [I]n truth there is only one mind."[74]

A similar premise has emerged from the work of researcher Roger Nelson, originally of the Princeton Engineering Anomalies Research (PEAR) lab, and his colleagues. They have examined the function of scores of random number generators situated around the globe. These electronic devices normally spit out patternless, equal numbers of ones and zeroes. But during moments when the attention of the world is riveted on a singular event, such as the death of Princess Diana or September

11, these mechanical devices deviate from their normally chaotic, random patterns and become more orderly. Nelson suggests that when the psyche of humans behaves collectively, it can impart order into situations where there was none.[75]

Whither?

It is easy enough to focus only on experimental findings that point to fundamental separations between biological entities. That is what our science has done for centuries, while denying any "unbroken wholeness" that may exist, as physicists Bohm and Hiley state in the epigraph.

A recurring rebuttal from the separateness camp is that any indication of unbroken wholeness is a temporary aberration based in faulty empiricism at best and fantasy at worst. When science is complete, this reasoning has it, any "science of connectedness" will yield to "science as usual"—the view of separate phenomena interacting through the customary local, physical forces recognized in contemporary physics and chemistry. Yet this is a faith-based view, because no one knows for certain what future developments may reveal. Science is open-ended and its accounts are never foreclosed. That is its strength, and that is what separates it from ideology. Nobel neurophysiologist Sir John Eccles and philosopher of science Karl Popper have called this ideology "promissory materialism"—the promise that one day science will give a complete description of the material basis for the whole of reality, including consciousness. Eccles: "Promissory materialism [is] a superstition without a rational foundation. [It] is simply a religious belief held by dogmatic materialists...who confuse their religion with their science. It has all the features of a messianic prophecy."[76]

If the emerging science of unbroken wholeness and nonlocal connectivity are incomplete, what of it? Incompleteness is a characteristic of the entire canon of science. All of science comes with a warning: "Until further notice." Uncertainty and incompleteness are necessary ingredients for better science. As mathematician and theoretical physicist Henri Poincaré stated, "Guessing before proving! Need I remind you that it is [through guessing] that all important discoveries have been made?"[77] In the same spirit, consciousness researcher Ian Stevenson, already mentioned, stated, "I believe it is better to learn what is probable about important matters than to be certain about trivial ones."[78]

The Ghastly Silence

For many individuals, the materialistic, intellectual formulations of science are not enough, because they omit too much of the juice of life. This deficiency in a purely scientific approach has long been noted by some of the greatest individuals in the history of science. Among them was Gottfried Wilhelm Leibniz (1646–1716), the German philosopher and mathematician. Leibniz, who invented the infinitesimal calculus independently of Isaac Newton, was considered one of the greatest minds of the eighteenth century. He refined the binary number system, which underlies virtually all digital computers, and invented mechanical calculators that were a marvel for their time. His intellectual reach touched all the major domains of learning of his day. Even so, Leibniz could not find within science the satisfaction he was looking for. In a letter two years before his death, he wrote:

> But when I looked for the ultimate reasons for mechanism, and even for the laws of motion, I was greatly surprised to see that they could not be found in mathematics but that I should have to return to metaphysics.[79]

Three centuries later, Erwin Schrödinger would come close to the same conclusion:

> The scientific picture of the real world around me is very deficient. It gives a lot of factual information, puts all our experience in a magnificently consistent order, but it is ghastly silent about all and sundry that is really near to our heart, that really matters to us. It cannot tell us a word about red and blue, bitter and sweet, physical pain and physical delight; it knows nothing of beautiful and ugly, good or bad, God and eternity. Science sometimes pretends to answer questions in these domains, but the answers are very often so silly that we are not inclined to take them seriously.[80]

The great Darwin also encountered the effects of the "ghastly silence" Schrödinger spoke of. Late in life he lamented, "My mind seems to have become a machine for grinding general laws out of large collections of facts.... The loss of [the emotional] tastes is a loss of happiness, and may possibly be injurious to the intellect, and more probably to the moral character, by enfeebling the emotional part of our nature.... The loss of these tastes is a loss of happiness." His solution: "[I]f I had to live my life again, I would have made a rule to read some poetry and listen to some music at least once every week...."[81]

Something more is needed—something that can marshal not only an intellectual appreciation of the wholeness implied in biological entanglement and nonlocality, but also something that can quicken the pulse and stir an ethic toward the earth that can counter the unbridled greed, selfishness and plunder that threaten us.

Currently there are excellent exemplars of this awakening, including numerous scientists. But many scientists, it must be said, are reluctant to speak out in favor of wholeness, unity, and oneness because they fear being labeled as having "gone mystic." It's as if there are hooded inquisitors lurking within science who are keeping score, and who are continually oiling the rack and heating the pincers, just waiting for a scientist to step out of line.

Fear has never silenced the greatest poets and artists, however. Poets have been yammering away about wholeness for centuries. As author Philip Goldberg points out in his important book *American Veda*,[82] there are superb examples among the Romantic poets, particularly William Blake, Percy Bysshe Shelley, William Wordsworth, and Samuel Taylor Coleridge. These poets sensed the interconnectedness and unity that are a feature of an entangled, nonlocal world. Thus Blake, in "Augeries of Innocence": "To see a world in a grain of sand / And a heaven in a wild flower, / Hold infinity in the palm of your hand / And eternity in an hour."[83] Shelley, in "Adonais": "The One remains, the many change and pass...."[84] Wordsworth, in "Tintern Abbey": "A motion and a spirit, that impels / All thinking things, all objects of all thought, / And rolls through all things."[85] And Coleridge, who wrote of "the translucence of the eternal through and in the temporal."[86]

In his book *Opening to the Infinite*, consciousness researcher Stephan A. Schwartz describes how the personal experience of a nonlocal event can carry the emotional wallop of an epiphany. Schwartz, who practically invented the science of remote viewing, has taught thousands of individuals in workshops to have these experiences. He concludes that nonlocal experiences, of which remote viewing is only one example, bestow an "ineffable sense of connection" and a "sense of empowerment" that is so profound it can permanently and radically alter one's worldview and conduct.[87]

The felt experience of being nonlocally connected—all tangled up with all there is—may be a way out of the mess created by self-centered, greed-obsessed individuals who have no sense of wholeness and no concern for the integrity of the earth. As Goldberg puts it, when we realize the unitary nature of consciousness,

> ...one's sense of "I" and "we" opens out from the narrow identification with family, tribe, race, political affiliation, religion, and so on, to encompass a broader swath of humanity. With that comes a corresponding expansion of the moral compass. This not a fanciful imagining of "we are the world" harmony but a living experience of unity with other humans, with nature, and ultimately with the cosmos.[88]

Straight-laced, paid-up scientists often deny the empirical findings pointing to an "unbroken wholeness" and unity between biological systems and humans, fearing the contamination of modern science by "the occult," one of their favorite epithets for nonlocal human experiences. But science desperately needs contamination by several factors that are missing from its equations, if we are to survive in any meaningful way. Some sort of connectivity is required for a moral center, an earth ethic, a sense of responsibility for all of life. The absence of these qualities has led to an abyss that is becoming impossible to ignore. A one-sided science is not only incomplete, it can be deadly. As Dr. Samuel Johnson put it nearly three centuries ago, "Integrity without knowledge is weak and useless, and knowledge without integrity is dangerous and dreadful."[89]

Dr. Johnson also observed, "When a man knows he is to be hanged in a fortnight, it concentrates his mind wonderfully."[90] Perhaps our sense of impending global disasters—I won't enumerate them—is concentrating our collective mind as a species, resulting in the return of ancient wisdom in the form of modern scientific insights, of which biological entanglement and nonlocality are an urgent example.

What we commonly call empathy, compassion, and love may be human entanglement banging on the doors of consciousness to gain entry. Albert Schweitzer, the legendary physician, missionary, priest, philanthropist, theologian, pacifist, musicologist, and winner of the 1952 Nobel Peace Prize, is an example of someone who opened those doors, and in so doing made the world a better place. In a kind of manifesto of wholeness, he wrote:

What we call love is in its essence Reverence for Life[91].... Profound love demands a deep conception and out of this develops reverence for the mystery of life. It brings us close to all beings. To the poorest and smallest, as well as all others.... [T]he idea of Reverence for Life gives us something more profound and mightier than the idea of humanism. It includes all living beings.[92]

Many consciousness researchers believe we are well on our way toward a revisioning of consciousness that is a reversal of the materialist views that have dominated science for generations. The emerging view is that consciousness is fundamental in its own right and is not produced by the physical brain; that consciousness may be the primordial organizing force of the universe and of life itself; and that the biological connectivity that we've examined above is a natural expression of consciousness that helps sustain life as we know it. Consider, for example, this observation in the 2017 book *Transcendent Mind: Rethinking the Science of Consciousness*:

We are in the midst of a sea change. Receding from view is materialism, whereby physical phenomena are assumed to be primary and consciousness is regarded as secondary. Approaching our sights is a complete reversal of perspective. According to this alternative view, consciousness is primary and the physical is secondary. In other words, materialism is receding and giving way to ideas about reality in which consciousness plays a key role.[93]

The authors of *Transcendent Mind*—Imants Barušs, PhD, professor of psychology at King's University College at Western University Canada, and Julia Mossbridge, PhD, experimental psychologist and cognitive neuroscientist at the Institute of Noetic Sciences and Visiting Scholar in Psychology at Northwestern University—contend that "the deep structures underlying our waking consciousness are fundamentally spatially and temporally nonlocal in nature." They explore empirical data, too long ignored, indicating that "consciousness is capable of existing in an extended or transcendent state in which it is not completely bound to the brain." This data supports the concept of shared mind, minds linked across space and time to form a collective, unitary human consciousness.[94]

Not only is the view of nonlocal, shared, transcendent, and fundamental mind supported by abundant empirical evidence, but these ideas also have an impressive pedigree. Max Planck, the founder of quantum mechanics, observed, "I regard consciousness as fundamental. I regard matter as derivative from consciousness. We cannot get behind consciousness. Everything that we talk about, everything that we regard as existing, postulates consciousness."[95] Erwin Schrödinger agreed: "Although I think that life may be the result of an accident, I do not think that of consciousness. Consciousness cannot be accounted for in physical terms. For consciousness is absolutely fundamental. It cannot be accounted for in terms of anything else."[96] More recently, mathematician-philosopher David Chalmers states, "I propose that conscious experience be considered a fundamental feature, irreducible to anything more basic.... "[97] And neuroscientist Christof Koch: "I believe that consciousness is a fundamental, an elementary, property of living matter. It can't be derived from anything else."[98] As to the concept of shared, unitary minds, we again find Schrödinger in agreement: "The overall number of minds is just one.... In truth there is only one mind." And as the eminent physicist

David Bohm observed, "Deep down the consciousness of mankind is one. This is a virtual certainty ... and if we don't see this it's because we are blinding ourselves to it."[99]

At this stage of humankind's existence, perhaps the best we can wish for one another is that we each simply realize that we're intimately connected with each other and everything through nonlocal, unitary, fundamental consciousness, and that we find the courage to allow this realization to make a difference in how we live our life. On this recognition, our future may depend.

Endnotes

1 Bohm D, Hiley BJ. *The Undivided Universe*. Reprint edition. London, UK: Routledge; 1995: 389.

2 Gronowicz GA, Jhaveri A, Clarke LW, Aronow MS, Smith TH. Therapeutic Touch stimulates the proliferation of human cells in culture. *The Journal of Alternative and Complementary Medicine*. April 1, 2008, 14(3): 233–239. doi:10.1089/acm.2007.7163.

3 Bengston WF. Spirituality, connection, and healing with intent: reflections on cancer experiments on laboratory mice. *The Oxford Handbook of Psychology and Spirituality*. (Lisa J. Miller, ed.). New York, NY: Oxford University Press; 2012: 548–577.

4 Bengston WF, Krinsley D. The effect of the laying-on of hands on transplanted breast cancer in mice. *Journal of Scientific Exploration*. 2000; 14(3): 353–364.

5 Bengston WF, Moga M. Resonance, placebo effects, and type II errors: some implications from healing research for experimental methods. *Journal of Alternative and Complementary Medicine*. 2007; 13(3): 317–327

6 Bengston, W. *The Energy Cure: Unraveling the Mystery of Hands-on Healing*. Sounds True Publishing; 2010.

7 Benor DJ. *Healing Research*. Vol. 1. Southfield, MI: Vision; 2002.

8 Jonas WB, Crawford CC. *Healing, Intention and Energy Medicine*. New York, NY: Churchill Livingstone; 2003: xv–xix.

9 Dossey L. *Reinventing Medicine*. San Francisco, CA: HarperSanFrancisco; 1999: 37–84.

10 Kelly EF, Kelly EW, Crabtree A, Gauld A, Grosso M, Greyson B. *Irreducible Mind: Toward a Psychology for the 21st Century*. Lanham, MD: Rowman and Littlefield; 2007.

11 Kelly EF, Crabtree A, Marshall P (eds.). *Beyond Physicalism: Toward Reconciliation of Science and Spirituality*. Lanham, MD: Rowman & Littlefield; 2015:

12 Schwartz SA. *Opening to the Infinite: The Art and Science of Nonlocal Awareness*. Buda, Texas: Nemoseen; 2007.

13 Schwartz SA, Dossey L. Nonlocality, intention, and observer effects in healing studies: laying a foundation for the future. *Explore (NY)*. 2010; 6(5): 295–307.

14 Radin D. *The Conscious Universe*. San Francisco: HarperSanFrancisco; 1997.

15 Radin D. *Entangled Minds*. New York, NY: Paraview/Simon & Schuster; 2006.

16 Bengston WF, Krinsley D. The effect of the "laying on of hands" on transplanted breast cancer in mice. *Journal of Scientific Exploration*. 2000;14(3):353–364.

17 Bengston W. *The Energy Cure: Unraveling the Mystery of Hands-on Healing*. Louisville, CO: Sounds True Publishing; 2010.

18 Sheldrake R. *Dogs That Know When Their Owners Are Coming Home: And Other Unexplained Powers of Animals*. New York, NY: Crown: 1999.

19 Roe CA, Sonnex C, Roxburgh E. Two meta-analyses of noncontact healing studies. *Explore*. 2015; 11(1): 11–23. Published Online at www.explorejournal.com: October 22, 2014. DOI: http://dx.doi.org/10.1016/j.explore.2014.10.001.

20 McMillan FD. The placebo effect in animals. *J Am Vet Med Assoc*. 1999;215(7):992–9.

21 Ader R, Cohen N. Behaviorally conditioned immunosuppression and murine systemic lupus erythematosus. *Science*. 1982; 215(4539): 1534–36.

22 Siegel S. Explanatory mechanisms for placebo effects: Pavlovian conditioning. In: *The Science of the Placebo: Toward an Interdisciplinary Research Agenda*. (H.A. Guess, ed.) London, UK: BMJ Books: 133–157.

23 Dossey L. Telecebo: Beyond placebo to an expanded concept of healing. *Explore*. 2015; 12 (1): 1–12.

24 Pizzi R, Fantasia A, Gelain F, Rossetti D, Vescovi A. Nonlocal correlation between separated human neural networks. In: Donkor E, Pirick AR, Brandt HE (eds.) *Quantum Information and Computation II*. Proceedings of SPIE5436. 2004:107–117. Abstract available at: The Smithsonian/NASA Astrophysics Data System. http://adsabs.harvard.edu/abs/2004SPIE.5436..107P. Accessed January 17, 2011.

25 Farhadi A, Forsyth C, Banan A, Shaikh M, Engen P, Fields JZ, Keshavarzian A. Evidence for non-chemical, non-electrical intercellular signaling in intestinal epithelial cells. *Bioelectro-chemistry.* 2007; 71 (2): 142–148.

26 Chaban VV, Cho T, Reid CB, Norris KC. Physically disconnected non-diffusable cell-to-cell communication between neuroblastoma SH-SY5Y and DRG sensory neurons. *Am. J. Translational Research.* 2013; 5(1): 69–79.

27 Duane TD, Behrendt T. Extrasensory electroencephalographic induction between identical twins. *Science.* 1965; 150(3694): 367.

28 Hearne K. Visually evoked responses and ESP. *Journal of the Society for Psychical Research.* 1977; 49, 648–657.

29 Hearne K. Visually evoked responses and ESP: Failure to replicate previous findings. *Journal of the Society for Psychical Research.* 1981; 51: 145–147.

30 Kelly EF, Lenz J. EEG changes correlated with a remote stroboscopic stimulus: A preliminary study. In: J. Morris, W. Roll, R. Morris (eds.). *Research in Parapsychology 1975.* Metuchen, NJ: Scarecrow Press; 1975: 58–63 (abstracted in: *Journal of Parapsychology.* 1975; 39: 25.

31 Lloyd DH. Objective events in the brain correlating with psychic phenomena. *New Horizons.* 1973; 1: 69–75.

32 May EC, Targ R, Puthoff HE. EEG correlates to remote light flashes under conditions of sensory shielding. In: Charles Tart, Hal E. Puthoff, Russell Targ (eds.). *Mind at Large: IEEE Symposia on the Nature of Extrasensory Perception.* Charlottesville, VA: Hampton Roads Publishing Company: 1979.

33 Millar B. An attempted validation of the "Lloyd effect." In: J. D. Morris, W. G. Roll, R. L. Morris (eds.). *Research in Parapsychology 1975.* Metuchen, NJ: Scarecrow Press; 1975: 25–27.

34 Millay J. *Multidimensional Mind: Remote Viewing in Hyperspace.* Berkeley, CA: North Atlantic Books; 2000.

35 Orme-Johnso, Dillbeck MC, Wallace K, Landrith GS. Intersubject EEG coherence: Is consciousness a field? *International Journal of Neuroscience.* 1982; (16): 203–209.

36 Rebert CS, Turner A. EEG spectrum analysis techniques applied to the problem of psi phenomena. *Behavioral Neuropsychiatry.* 1974; (6): 18–24.

37 Targ R, Puthoff H. Information transmission under conditions of sensory shielding. *Nature.* 1974; (252): 602–607.

38 Grinberg-Zylberbaum J, Ramos J. Patterns of interhemispheric correlation during human communication. *International Journal of Neuroscience,* 1987: (36): 41–53.

39 Grinberg-Zylberbaum J, Delaflor M, Attie L. The Einstein-Podolsky-Rosen paradox in the brain: The transferred potential. *Physics Essays.* 1994: (7):422–428.

40 Grinberg-Zylberbaum J, Delaflor M, Sanchez ME, Guevara MA. Human communication and the electrophysiological activity of the brain. *Subtle Energies and Energy Medicine.* 1993; 3: 25–43.

41 Sabell A, Clarke C, Fenwick P. Inter-Subject EEG correlations at a distance — the transferred potential. *Proceedings of the 44th Annual Convention of the Parapsychological Association.* New York, NY: Parapsychological Association; 2001: 419–422.

42 Standish L, Kozak L, Johnson LC, Richards T. Electroencephalographic evidence of correlated event-related signals between the brains of spatially and sensory isolated human subjects. *Journal of Alternative and Complementary Medicine.* 2004: 10(2), 307–314.

43 Standish L, Johnson, LC, Richards T, Kozak L. Evidence of correlated functional MRI signals between distant human brains. *Alternative Therapies in Health and Medicine.* 2003: (9): 122–128.

44 Wackerman J, Seiter C, Keibel H, Walach H. Correlations between brain electrical activities of two spatially separated human subjects. *Neuroscience Letters.* 2003; (336): 60–64.

45 Standish L, Johnson LC, Richards T, Kozak L. Evidence of correlated functional MRI signals between distant human brains. *Alternative Therapies in Health and Medicine.* 2003: (9): 122–128.

46 Standish L, Kozak L, Johnson LC, Richards T. Electroencephalographic evidence of correlated event-related signals between the brains of spatially and sensory isolated human subjects. *J. Alternative and Complementary Medicine.* 2004: 10(2), 307–314.

47 Kittenis M, Caryl P, Stevens P. Distant psychophysiological interaction effects between related and unrelated participants. *Proceedings of the Parapsychological Association Convention 2004*: 67–76. Meeting held in Vienna, Austria, August 5–8, 2004.

48 Radin D. Event-related electroencephalographic correlations between isolated human subjects. *Journal of Alternative and Complementary Medicine*. 2004; (10): 315–323.

49 Fowler JH, Christakis NA. Dynamic spread of happiness in a large social network: longitudinal analysis over 20 years in the Framingham Heart Study. *British Medical Journal*. 2008; 337: a2338.

50 Belluck P. Strangers may cheer you up, study shows. New York Times online. http://www.nytimes.com/2008/12/05/health/05happy-web.html. December 4, 2008. Accessed January 18, 2009.

51 Stein R. Happiness can spread among people like a contagion, study indicates. Washington Post online. http://www.washingtonpost.com/wp-dyn/content/story/2008/12/04/ST2008120403608.html. December 5, 2009. Accessed January 18, 2009.

52 Bond M. Three degrees of contagion. *New Scientist*. 2009; 201 (2689): 24–27.

53 Christakis NA, Fowler JH. *Connected: The Surprising Power of Our Social Networks and How They Shape Our Lives*. Boston, MA: Little, Brown and Company; 2009.

54 Christakis NA, Fowler JH. The spread of obesity in a large social network over 32 years. *New England Journal of Medicine*. 2007; 357: 370–379.

55 Kaplan K. Happiness is contagious, research finds. Los Angeles Times online. http://articles.latimes.com/2008/dec/05/science/sci-happy5. December 5, 2008. Accessed January 19, 2009.

56 Bobrow RS. Evidence for a communal consciousness. *Explore: The Journal of Science and Healing*. 2011; 7(4): 246–248.

57 Priestley JB. *Man & Time*. London, UK: W. H. Allen; 1978: 211–212.

58 Schwarz BE. Possible telesomatic reactions. *The Journal of the Medical Society of New Jersey*. 1967;64(11):600–3.

59 Gurney E, Myers, FWH, Podmore F. *Phantasms of the Living*. Volume 1. London: Trübner; 1886: 188–189.

60 Lorimer D. *Whole in One*. London: Arkana/Penguin; 1990: 72–105.

61 Stevenson I. *Telepathic Impressions: A Review of 35 New Cases*. Charlottesville, VA: University Press of Virginia; 1970.

62 Rush JH. New directions in parapsychological research. *Parapsychological Monographs No. 4*. New York: Parapsychological Foundation; 1964:18–19.

63 Rhine LE. Psychological processes in ESP experiences. Part I. Waking experiences. *Journal of Parapsychology*. 1962; 29: 88–111.

64 Playfair GL. *Twin Telepathy: The Psychic Connection*. London, UK: Vega; 2002.

65 Playfair GL. *Twin Telepathy: The Psychic Connection*. London, UK: Vega; 2002: 12.

66 Playfair GL. *Twin Telepathy: The Psychic Connection*. London, UK: Vega; 2002: 16.

67 Guy Lyon Playfair. *Twin Telepathy: The Psychic Connection*. London, UK: Vega; 2002: 51.

68 Dossey L. *Healing Words*. San Francisco, CA: HarperSanFrancisco; 1993: 39–44.

69 Nadeau R, Kafatos M. *The Nonlocal Universe: The New Physics and Matters of the Mind*. New York, NY: Oxford University Press; 1999.

70 Herbert N. *Quantum Reality*. Garden City, NY: Anchor/Doubleday; 1987: 214.

71 Feather SR, Schmickler M. *The Gift: ESP, the Extraordinary Experiences of Ordinary People*. New York: St. Martin's Press; 2005:2.

72 Cox WE. Precognition: An analysis II. *Journal of the American Society for Psychical Research*. 1956; 50 (1): 99–109.

73 Schrödinger E. *My View of the World*. Woodbridge, CT: Ox Bow Press; 1983: 31.

74 Schrödinger E. *What Is Life? and Mind and Matter*. London, UK: Cambridge University Press; 1969: 139.

75 , 2002; 15(6): 537–550.

76 Eccles J, Robinson DN. *The Wonder of Being Human*. Boston: Shambhala; 1985:36.

77 Poincaré H. Quoted in: *La valeur de la science*. In Anton Bovier, *Statistical Mechanics of Disordered Systems*. Cambridge, UK: Cambridge University Press. 2006: 218.

78 Stevenson I. *Reincarnation and Biology.* Westport, CT: Praeger; 1997: 186

79 Leibniz GW. Quoted in: Stanford Encyclopedia of Philosophy online. Gottfried Wilhelm Leibniz. http://plato.stanford.edu/entries/leibniz/. Accessed July 20, 2011.

80 Schrödinger E. Quoted in: *Quantum Questions* (Ken Wilber, ed.). Boulder, CO: New Science Library; 198 : 81

81 Darwin C. Quoted in: *The Life and Letters of Charles Darwin.* Vol. 1. (F. Darwin, ed.) New York; D. Appleton & Co.; 1897: 81–82.

82 Goldberg P. *American Veda.* New York, NY: Harmony; 2010: 270.

83 Blake W. From: Augeries of Innocence. Quoted in: Bartlett, John. *Bartlett's Familiar Quotations* (Justin Kaplan, general ed.). Sixteenth Edition. Boston: Little, Brown and Company; 1992:359.

84 Shelley PB. From: Adonais. Quoted in: Bartlett, John. *Bartlett's Familiar Quotations* (Justin Kaplan, general ed.). Sixteenth Edition. Boston: Little, Brown and Company; 1992:409.

85 Wordsworth W. From: Tintern Abbey. Quoted in: Bartlett, John. *Bartlett's Familiar Quotations* (Justin Kaplan, general ed.). Sixteenth Edition. Boston: Little, Brown and Company; 1992:373.

86 Coleridge ST. *The Statesman's Manual: Critical Theory Since Plato.* (Hazard Adams, ed.) New York, NY: Harcourt Brace Jovanovich; 1971: 476.

87 Schwartz SA. *Opening to the Infinite: The Art and Science of Nonlocal Awareness.* Buda, Texas: Nemoseen; 2007: 38.

88 Goldberg P. *American Veda.* New York, NY: Harmony; 2010:346.

89 Johnson S. Quoted in: Quoteworld.com. http://www.quoteworld.org/quotes/7290. Accessed July 24, 2011.

90 Johnson S. Quoted in: Quoteworld.com. http://www.quoteworld.org/quotes/7290. Accessed July 24, 2011.

91 Schweitzer A. *Indian Thought and Its Development.* (Mrs. Charles E. B. Russell, trans.) New York, NY: Beacon Press; 1934: 260.

92 Schweitzer A. Wikiquote: Albert Schweitzer. http://en.wikiquote.org/wiki/Albert_Schweitzer. Accessed July 12, 2011.

93 Barušs I, Mossbridge J. *Transcendent Mind: Rethinking the Science of Consciousness.* Washington, DC: American Psychological Association; 2017: 3.

94 Barušs I, Mossbridge J. *Transcendent Mind: Rethinking the Science of Consciousness.* Washington, DC: American Psychological Association; 2017: 81, 171.

95 Planck M. *The Observer.* London, UK; January 29, 1931.

96 Schrödinger E. Quoted in: Walter Moore. *A Life of Erwin Schrödinger.* Canto edition. Cambridge, UK: Cambridge University Press. 1994: 181.

97 Chalmers DJ. The puzzle of conscious experience. *Scientific American.* 1995; 273(6): 80–6. http://consc.net/papers/puzzle.pdf. Accessed 20 November, 2016.

98 Koch C. *Consciousness: Confessions of a Romantic Reductionist.* Cambridge, MA: MIT Press. 2012: 119.

99 Bohm D. Quoted in: Renée Weber. *Dialogues with Scientists and Sages.* New York, NY: Routledge & Kegan Paul; 1986: 41.

Complexity, Complementarity, Consciousness

VASILEIOS BASIOS

Several modern scientific disciples are rapidly exhausting the one-sided mechanical and reductionistic thinking upon which they were established. Biological Evolution is discussed as such an example here. When confronted with the complexities of reality, our ideas about biological evolution had to change tenets and seek new grounds for its foundations. The complementarity of function and structure, as principle and as phenomenon, is elaborated further to help us discover how unity sustains dualities and why complexity arises unavoidably from polarities. Complementarity and Complexity, ubiquitous as they are, point to the need for a new kind of scientific endeavor that simultaneously brings forth and is brought from a deeper understanding of the workings of Consciousness. From the example of the science of evolution we can see the next turn of the evolution of science. To this end we trace some novel realizations about the role of complementarity and complexity in logic, neurosciences, psychology, and philosophy. They all point to that need for a new kind of understanding and a new kind of science. This can only be a "science towards the origins." And as its origin is consciousness, we realize that for this new kind of science to emerge, a new kind of consciousness has also to emerge in parallel. So, we attempt to propose a coarse outline for their new alliance.

Prologue: Evolution is not what it used to be...

There was a time when the linear mechanistic view of evolution dominated the mind and practices of the academic and research communities all over the world. After the deciphering of the "alphabet of life," as the genetic code of Crick and Watson came to be known, the Central Dogma of molecular biology became just that! A powerful unquestionable dogma that dictated the program of biology. It blandly stated that the chain of command in life everywhere was going from the molecular level up. The official statement proclaimed by Crick was "DNA makes RNA and RNA makes protein." The idea was that a linear flow of information establishes a strict hierarchy of functions totally dependent and subservient to the structure of the DNA macromolecules. Copying the dominant idea of determinism that shaped classical physics up to the last century, biology followed the dream of establishing a grand project of understanding and controlling life by understanding and controlling its structure: the omniscient and omnipotent molecule of DNA! The gene and the DNA became the elementary particles of biology. So biology's final task was to discover the "alphabet, language and logic of life" by deploying the grand "Human Genome Project." The genes even took anthropomorphic qualities, like "selfish," "intelligent," and "virtuous." They were held responsible not only for our diseases but also for our careers, vices, virtues, and even religions and god. In a way, they were collectively considered as the new immortals ruling over humankind.

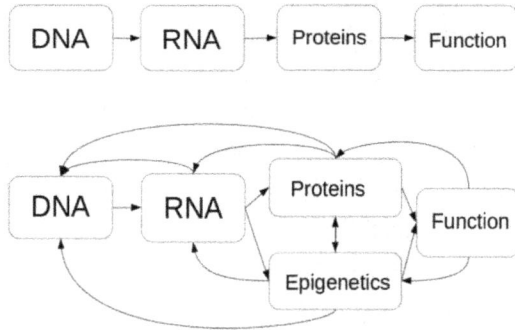

Figure 1 – Life's complications…information flow is non-linear

Alas for the central dogma, eventually the complexities of life took over. The idea that the DNA drives evolution only through random mutation was deflated and abandoned. We now know that the number of genes do not reflect the differences between humans and other organisms. The verdict was out with a funny surprise for the naive mechanistic/reductionist mind. The human genome was found to consist of only about twice the number needed to make a fruit fly, worm, or plant![1] Phenomena like "alternative splicing" and other complicated feedback processes finally drew the attention that was lacking under the spell of the "central dogma." For example and quite interestingly, prions, the notorious protein pathogens of "mad-cow" disease, have been recently discovered to be genetic elements that store and transmit information in various organisms.[2] It has now been established that are many more ways in which a gene's protein-coding sections (exons) can be joined together to eventually create a functional protein. Indeed, now "we cannot escape the conclusion that physical and behavioural differences between species are not related in any simple way to gene number."[1] The complexity of the evolutionary processes cannot be reduced to simple molecular mechanisms driven by pure randomness.

To his credit, the great developmental biologist Richard Strohman had predicted these surprises well in advance of the "Human Genome Project" completion. He foresaw the important functional role in evolution of the various feedback processes to genome from epigenetic factors. He brilliantly argued about the need to contain the paradigm of the gene, and for that matter of all paradigms, to its rightful place. He warned us, as early as 1997, about the unscientific slippery slope of "Big Science." As he provocatively put it, "according to all media reports, genetic determinism is a paradigm whose time is here and now: everyone will get better as their biotherapists become richer."[3] His prediction for the then upcoming limits of the genetic paradigm and his brilliant deconstructing of the "myth of the gene" that aids the prevailing naive ideas of evolution led him to consider alternatives to understanding evolutionary processes in the light of ideas close to Waddington and Wallace. Ideas that call for understanding the inter-woven phenomena of genetic-epigenetic interactions in the light of complexity, self-organization, adaptive systems, and emerging patterns and processes.[3,4]

It is customary to attribute to Darwinism the foundation of molecular biology and to use genetics to "prove" Darwinism. But actually, Darwin himself was not an anti-Lamarckian. With his friend and co-founder of the idea of evolution through adaptation, the lesser-known pioneer Alfred Russel Wallace, he left the possibility of adaptation due to environmental interactions open. Although Darwinists managed to prove Lamarckians wrong, contemporary developments towards the "extended evolutionary synthesis" bring back again the ideas of the two protagonists as they were at the time when they were jointly proposing their theory of the origins of

the species. Wallace, one of the most talented and potent intellectuals of
the late nineteenth century, went even further than Darwin in proposing
the intervention of "a mind of the species" as a leading factor of evolution. No wonder why the original "Wallace-Darwin Theory" changed to
the "Darwin-Wallace Theory" and then just morphed down to what now
is known as "Darwin's Theory." (A detailed account of the erasing from
the collective academic memory of Wallace's name as well as his brilliant
life and legacy can be found in *Natural Selection and Beyond.*[5])

Although Wallace was punished for his "radical" socioeconomic
ideas and his fondness for spiritualism, other prominent proposals about
a not-so-random, even guided, evolution were preparing the "extended
evolutionary synthesis." A modern synthesis leads us one step closer to
revisiting Bergson's ideas of "Creative Evolution."[6] His criticism equally
of both the mechanistic approach of genetic determinism as well as teleological finalism (i.e. that everything is "designed" to evolve as it evolves)
still holds true as a valid ground of understanding change and emergence
as a self-organizing process: Life equals creativity. Bergson's "vital impulse" puts the "telos" of life at the very origin, the "initial conditions" in
today's terminology. Equally important is the phenomenon of life's "com-
plexification." This is the fact that life goes on creating and evolving from
the simpler to the more complex organisms. Simple procaryotic cells will
give rise to eukaryotes, eukaryotes make more complex cells and even
form big cellular cooperatives as they give rise to fungi and plants, and as
life goes on it differentiates the species so that they are made of more and
more complex individuals in more and more complex interrelationships.
In fact, it is worth observing that as life evolves and complexifies the degrees of freedom thus afforded increase. The cells diffuse and are swept
away, the plants grow, the animals move, the more developed animals acquire complex brains that think and control their environment. If we want
to put humans on the apex evolution we can even dare say they control
completely their environment, albeit they still have to demonstrate that
they can do that in a sustainable way.

If we follow Bergson's lead, life must be equated with creation. The
coexistence of automated instinct, mostly "unconscious," and purposeful
intelligence, mostly "conscious," are two complementary poles and not
mutually exclusive states. They both stem from what informs life itself:

change and becoming. In that sense, we have to think again of what we mean by the dualities chance (randomness) and order (law). As Bergson asks "order is certainly contingent, but in relation to what?" to answer that it is not a matter of order versus disorder, but rather of one order, or pattern, in relation to another.

Structure & Function, Objects & Processes: Complementarity is Life's Force

From the "one-way only" idea that genetic material's (DNA) molecular structure and random mutations would explain in a grand scheme the totality of biological function, life, and its evolution we have now arrived at the inescapable conclusion that life's function itself equally determines its own genetic structure. Through the vicissitudes of modern times, the complexities of gene expression, epigenetic regulations, environmental/ecological pressures, and a plethora of other factors, we have come to understand that structure and function cannot each work independently and in separation. More and more we now seek their harmonious coexistence. The point that offers us an understanding is a point at the correct place in the continuum these contraries define. Complementarity is the key idea here.

Complementarity usually brings forth the ideas of its primary advocate and one of the founders of quantum physics, Niels Bohr. His dictum, which he also chose for his coat of arms, "*contraria sunt complementa*" (opposites are complementary) along with his chosen picture of Tao's yin-yang poles has been stirring up excitement and controversy since the early days of quantum mechanics. Complementarity a la Bohr has been hailed as a great revolution in modern thinking, a daring transcendence, which allows understanding of the wave-particle nature of light and every other elementary particle thereafter discovered.

For physics, complementarity is a theoretical principle as well as an established experimental fact. The Complementarity principle states that properties that cannot all be observed or measured simultaneously, coexist in

Bohr's coat of arms

a complementary fashion. Bohr considered as complementary dualities the fundamental, mutually exclusive quantum properties. In particular, "position and momentum," "energy and duration," "spin on different axes," "wave and particle," even the value of a field and its local change and the equally celebrated ever since, key properties of quantum systems: the duality of "entanglement and coherence." The above quantum properties of matter are mutually exclusive due to the celebrated "Uncertainty Principle" put forth by Werner Heisenberg, whose uncertainty relations are the sine qua non characteristic of any quantum theory.

Concepts like "particle" and "wave," which are clearly borrowed from classical physics, make it impossible for an object to be particle and wave at the same time. So, Bohr argues, it is impossible to fully measure wave and particle aspects simultaneously. In quantum mechanics, intrinsic properties are dependent of their determination by a measuring device. This is a strong statement but also an indisputable experimental fact. In relatively recent times this theoretical statement, supported by the Kochen-Specker theorem and the violation of Bell's inequalities, has been extensively tested in determining the nonlocality of entangled/coherent quantum states. So the verdict of quantum theory has been confirmed again and again that "the type of measurement determines which property is shown."[26]

This is all quite well known and widely discussed. What is not so well known is that Niels Bohr, and many other pioneers of quantum mechanics, strongly believed that complementarity has a wider area of application than quantum physics.[7,9,23,24] As his life and work progressed, he believed in the universality of his principle more and more strongly. Actually, the complementarity principle's roots lie deep in biology.

It is reported that as early as 1929 Bohr briefly noted that "in the description of living organisms one might see a certain connection with the issues of the quantum theory."[7] In an international congress of light therapists (!) in Copenhagen, circa 1932, he addressed the audience with his lecture titled *"Life and Light."* There he posed the question of whether or not the analysis of living processes could be reduced and described in terms of pure physical-chemical mechanisms. Interestingly enough, in 1972, Kurt Goedel, the greatest logician since Aristotle, also raised the same question. No surprise here since this question bears the most critical answer for the future development of science. Their common question was whether or not our physical and biochemical substratum permits a mechanical interpretation of all the functions of life and the mind. Bohr, although he emphasized the uniqueness of life in terms of organization (structure) and teleological purposefulness (function), he feared being blamed as an old fashioned vitalist and did not give a clear answer. As he put it: "If we were able to push the analysis of the mechanism of living organisms as far as that of atomic phenomena, we should scarcely expect to find any features differing from the properties of inorganic matter." So, he left the scholars divided. Did he reduce life down to quantum mechanisms or did he push teleology and purpose down to the quantum level?

Kurt Goedel, on the other hand, touched upon on the nature of consciousness, life, and mind in a more explicit way.[8] Indeed, any possible answer to this question hinges upon how complexity emerges in quantum systems, the borderline between quantum theory and biochemistry, and in the application of algorithmic complexity theory to quantum information theory. What is at stake here requires an interdisciplinary effort of immense proportions, so we leave it for the present noting that, indeed, how we understand life and mind (whether these are cosmic phenomena or mere earthbound accidents) depends on the outcome of this question. But principally their outcome depends upon how we pose these questions.[8,23,24] Let us return then to the principle of complementarity beyond quantum physics.

Bohr extended his principle into biology by stating:

> The question at issue is whether some fundamental traits are still missing in the analysis of natural phenomena before we can reach an understanding of life on the basis of physical experience ... It must be kept in mind, however, that the conditions in biological and physical research are not directly comparable, since the necessity of keeping the object alive imposes a restriction on the former [i.e. living things] which finds no counterpart in the latter. Thus, we should doubtlessly kill an animal if we tried to carry the investigation of its organs so far that we could tell the part played by the single atoms in vital functions. In every experiment on living organisms there must remain some uncertainty as

regards the physical conditions to which they are subjected, and the idea suggests itself that the minimal freedom we must allow the organism will be just large enough to permit it, so to say, to hide its ultimate secrets from us. On this view, the very existence of life must in biology be considered as an elementary fact, just as in atomic physics the existence of the quantum of action has to be taken as a basic fact that cannot be derived from ordinary mechanical physics. Indeed, the essential non-analyzability of atomic stability in mechanical terms presents a close analogy to the impossibility of a physical or chemical explanation of the peculiar functions characteristic of life.[27]

And Bohr continues:

"[there is an]…obvious exclusiveness between such aspects of life as the self-preservation and self-generation of individuals on the one hand, and the subdivision necessary for any physical analysis on the other hand. Due just to this essential feature of complementarity, the concept of purpose which is foreign to mechanical analysis finds a certain application in biology."

As his biographer Abraham Pais reports, "later Bohr expressed his views most succinctly like this: 'Mechanistic and vitalistic arguments are used in a typically complementary manner.' "[7] Of course, such a statement ran counter to the current of ideas prevailing in biology at that time; no wonder that Bohr's lecture "Light and Life" was considered almost scandalous and its fate in citations was set from dim to obscure.

Is life's force behind the principle of complementarity, or is complementarity as a real phenomenon behind the concept of life's force? Or maybe this recurrent question simply occurs in unison? Let us keep our minds and questions open. It is remarkable that contemporary research on the history of the development of the idea of complementarity provides evidence that the principle of complementarity is not just a loan of biology from physics but that their inter-penetration is deeper than we thought. Niels Bohr's father, Christian Bohr, an eminent professor of physiology, took part in the debate whether the exchange of oxygen and carbon dioxide in the lungs (an important problem after the First World War's use of poisonous gas as weapons) could be explained as a diffusion process. His conclusion was that there are other regulatory processes and feedbacks that are governed by the needs of the organism to be sustained as a whole. Young Niels Bohr's mind was definitely influenced by these heated debates that he witnessed about the preeminence of structure or function between vitalists and mechanists,[9] and sparked his life-long interest in biology.

As physicist Basil Hiley never ceases to advocate, "For Bohr, this was an indication that the principle of complementarity, a principle that he had previously known to appear extensively in other intellectual disciplines but which did not appear in classical physics, should be adopted as a universal principle."[10] Hiley, an erudite scholar of quantum physics, mathematics and philosophy has followed the ideas of complementarity outside Bohr's Copenhagen interpretation of quantum mechanics. Hiley's work with David Bohm on the implicate/explicate order, or as they call it, "undivided wholeness," and the concept of "holomovement" exceed also the boundaries of physics to include matter and cosmos, life and consciousness.[11] Besides the brilliant development of the mathematical framework of the "holomovement," their ideas are grounded in a deep philosophical background that can be traced back to Baruch Spinoza, Gottfried Wilhelm Leibniz, Giordano Bruno, and Nicholas of Cusa (or Cusanus).[12]

When discussing the enfolding-unfolding processes in the universe and consciousness, David Bohm elaborated further on the need for new notions of order in physics and in science in general. According to Bohm and Hiley, totalities are continually forming and dissolving out of the universal flux, the holomovement. The different poles of dualities here are seen as not only as complementary, but as identical, stemming from a deeper undivided source. The implicate order unfolds to the explicate order of phenomena "in a state of unending flux of enfoldment and unfoldment, with laws most of which are only vaguely known." Although unobservable or even unspeakable, the implicate order is felt and real. It is indirectly detected in the emerging, explicate processes, in the phenomenal world. Bohm concludes: "All of these [matter, life, cosmos, and consciousness] have been considered to be projections of a common ground.

This we may call the ground of all that is." Others call it "The Source," "The Force," "The Godhead," or "Tao."

Both Bohm and Bohr, although the founders of different interpretations of quantum mechanics, see complementarity as the unifying, ubiquitous (both as a principle and as a phenomenon), common ground of all that is. They were both aware of the Hermetic-Alchemical intellectual tradition's idea of *"coniunctio,"* meaning conjunction, as the (al)chemical process where two chemical substances "marry" to produce a third, different chemical substance. Carl Gustav Jung used also this term in his psychoanalytical work to describe an unconscious experience (e.g. instinct) that is combined with consciousness and becomes a new different experience (e.g. desire). Wholeness requires a *"coniunctio oppositorum"* (conjunction of opposites), an alchemical marriage. Here contradictory aspects are not just complementary, they are identical under conjunction. The ends of any polarity meet in the implicate and appear as separate aspects in the explicate. The whole, total flux is all that is all around.

Linear and Non-linear Logics: Similar but Different, Different but Similar

May you be able to find the similarities in difference and the differences in similarity.
— THE MASTER OF THE DIAMOND (BY EMILIOS BOURATINOS)

In tristitia hilaris in hilaritate tristis (in sadness joyful, in joyfulness sad)
— GIORDANO BRUNO, "IL CANDELAIO"(THE CANDLEHOLDER)

Complementarity leads us to the strange logic of quantum physics with its seemingly contradictory, counterintuitive and sometimes upsetting conclusions for our everyday, classical, mindset. These "quantum paradoxes" caused John von Neumann and Garett Birkhoff to examine the logical foundations of quantum physics. And so they offered to the world another genuine surprise: quantum physics cannot be cast in a standard Aristotelean, or Boolean, framework. We need to extend our logical framework to accommodate complementarity and the weirdness of the quantum world. So, they proceeded in doing exactly that.

It generated a wide array of studies and now we understand that the logic of quantum physics can be formulated as a modified version of the standard propositional logic. There are many names and versions of "quantum logics" in our days, non-commutative logic, many-valued logic, non-associative logic, and the list can go on.

Self-reflecting Taijitu.
Yin in Yang in Yin...

Quantum logic is a new kind of logic; that is, a formal way to reach conclusions from premises (or presumptions). It has certain properties that differentiate it from classical (Aristotelean/ Boolean) logic. The most crucial distinction from classical logic is that the "distributive law" does not hold. In Aristotelean logic, as formulated in mathematical language by George Boole, if we have three propositions (say, A, B, C) their logical operations of conjunction ("and") and disjunction ("or") can be combined as: *"A and (B or C)" is equivalent to "(A and B) or (A and C)."* So, if one reads the menu that offers "eggs and bacon or sausages," if the restaurant honors Aristotelean logic one can safely order "eggs and bacon" or "eggs and sausages... unless the restaurant's waiters are quantum physicists! Because in a quantum restaurant that prides itself of its non-classical, quantum, logic if you see offered "eggs and bacon or sausages" and request "eggs and bacon" it will be entirely logical for them to inform you that they cannot serve you either "eggs and bacon" nor "eggs and sausages", because their "distributive law" is broken. However, you can have "eggs and (sausage or bacon)" where "(sausage or bacon)" will be a concoction of processed meat that resembles bacon and sausages at once...*

When the distributive law is not observed, the three tenets of classical, Aristotelian, logic also cannot hold unconditionally. These are, as Bertrand Russell defined them: *The law of identity*: "Whatever is, is." *The law of*

---

* This illustration of quantum logic appears in endnote 13 from where I modified it and transferred it here. Another friend, David Lorimer, when discussing the logic of quantum observables offers another instance: "A client asked for tea or coffee has just been served a brown beverage. After tasting it, horrified, asks the quantum waiter 'Oh, jees! Is this tea or coffee?' only to get the answer 'if you can't tell the difference why does it matter?' ... uncertainty and complementarity principles at work in quantum restaurants!"

contradiction: "Nothing can both be and not be." And *The law of excluded middle*: "Everything must either be or not be." Hence one can rightfully ask, what is the use of logic and what is reasonable in logic? Here one has to contemplate before rejecting any non-classical logic that, as in ordinary life, usually arise in situations in need of understanding, and requiring some reasoning that is far too complex and far too non-linear than the clear "yes-no" universe of classical Boolean/Aristotelean strict logic, that we demand from a good restaurant and on which the "Boolean lattice of propositions" our computers and machines are built upon.

As far as quantum physics is concerned, however, the obstacles to conventional logic are the uncertainty principle, the nature of entanglement, and the complementarity issues of incompatible measurements. Uncertainties can play an equally obstructing role to complex systems' descriptive logical attempts. As complexity aspects are malleable due to their multi-faceted nature and/or adaptable under different conditions, during the process of observation we expect that complexity would raise equally disconcerting questions about their "unreasonable" extended logic as viewed through our classical mindset.

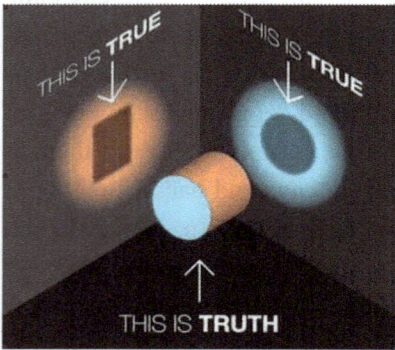

...what is true and what is truth?
(cc) by Leigh Blackall

Computers and machines do work based on Boolean logic, for sure. But it is exactly these severe limitations imposed by Boolean logic that ignited the search for extensions of classical logics in order for our machines to deal with the surrounding complexities of real world problem solving. Many models of alternative logics have been and continue to be proposed, sometimes inspired by quantum logic. The names of these logics are as innovative as their creators: we have Fuzzy logic, Rough-set logic, paraconsistent (or deviant) logics, many-valued logics, intuitionistic logic, and temporal logics, to name a few. Evidently, not all of them are directly related to quantum logic. Yet, context dependent logics and temporal logics, logics that change as time or information flow goes by, are extremely close and relevant to quantum logic.[13,23,24]

It has been understood that Aristotelean/Boolean logic is based on language and on object-mediated perception. Then these other fancy logics could be based on another kind of language, accessible and manipulable through their strict formalism and even special machine coding. Then the problem of truth arises. And this problem is as old as philosophy itself. The problem of truth took a major decisive turn after Socrates rebelled against sophists for their use of logic that could prove whatever they liked whenever it seemed profitable to their petty interests. Are we now witnessing a similar major shift in understanding of the problem of truth as then? What will guide us to truth now that proof can lead us potentially anywhere?

If logic can create monsters what can guide logic to truth? Henri Poincare, the father of modern chaos theory, insisted that "it is by logic that we prove, but by intuition that we discover;" and that "Logic teaches us that on such and such a road we are sure of not meeting an obstacle; it does not tell us which is the road that leads to the desired end. For this, it is necessary to see the end from afar, and the faculty which teaches us to see is intuition." It is exactly what the contemporary Greek philosopher and essayist Emilios Bouratinos has been advocating all along: a return to "Logos" as a guide to escape the inevitable irrationalities of any system of logic that traps our mind in any particular paradigmatic thinking. As he puts it,

> One of the important things modern science has revealed is that when we objectify things, there is a price to pay. Objectifications always end up with something less than the real thing itself. They lead to a conceptual crystallisation of entities which in fact have acquired only temporary form and structure. So we cannot know with absolute certainty what will and what will not change in them, when it will change and to what extent they will not change. The origins of strange attractors in the non-detectable initial circumstances of chaos theory, plus Goedel's incompleteness theorem and Heisenberg's uncertainty principle, render any sweeping generalisation about patterns, transformations, methods or outcomes unreliable.[19]

Socrates would have readily agreed with Poincare and Bouratinos, as many contemporary thinkers agree in trusting, and leading us in trusting too, the forgotten organ of intuition anew. This is the only hope in navigating out of the labyrinth our collective mindset traps us in. Such realizations escape from the confines of academic thinking; nowadays many think along the same lines. Mainly due to the increasing complexities that our modern civilization faces and its inability to resolve crucial issues that fast become matters of life and death, more and more institutions, thinkers and common folks aspire for another way of understanding and doing things. Robert Jahn and Brenda Dunne in their book *Consciousness and the Source of Reality*[14] quote a wonderful proverb carried forth from perennial wisdom by the Sufis, it goes: "You think that because you understand ONE you understand TWO, because one and one makes two. But you must understand AND!" It cannot be made more succinct or lucid than that. It stands so true that when dealing with logical reasoning we cannot be led blindly by assumptions, consciously or unconsciously given. We have to turn to honest introspection and be able to see the context in which our reasoning operates. It is not only a question of how, but of why, and from where, what is given arises. And of course, it is imperative to be able to discern the given from the real, in each and every case, at each and every time.

The One Behind the Two

There are powers and thoughts within us, that we know not till they rise
Through the stream of conscious action from where the self in secret lies.
— JAMES CLERK MAXWELL, in
The Man Who Changed Everything[22]

The faculty of intuition is not something that speaks only to the poet or the artist. It is inherent in all of us even though latent, silent or unacknowledged. To be able to access it as the great scientists have always done might actually be easier than we think. As proof has to be guided by truth in order to be successful, intuition has to be guided by beauty and utility by what is pleasurable and good. Let us call Henri Poincare again on the stage, he says: "The scientist does not study nature because it is useful to do so. He studies it because he takes pleasure in it, and he takes pleasure in it because it is beautiful. If nature were not beautiful it would not be worth knowing, and life would not be worth living.... What I mean is this more intimate beauty which comes from the harmonious order of its parts, and which a pure intelligence can grasp."[28]

There is where the self in secret lies, close to the source of beauty and truth. It is where the "coinsidentia oppositorum" (coincidence of opposites) of Nicholas of Cusa, the complementarity of Bohr, and the implicate order of Bohm meet with no paradox. We can see and feel that this is so for all dualities, either confronted or created. We can find endless examples. For one, the crux of understanding Complexity, which is the duality of parts and whole. We have come to understand—especially with the recent advent of chaos and complexity theory, nonlinear dynamics, self-organization and systems' science—that indeed "the whole is more than the sum its parts" or that "more is different." What remains to be kept in mind along with these often-quoted remarks is that indeed the whole is reflected in all the parts, that all things keep their own relationship to the whole as the whole interpenetrates its parts. So the dichotomies of parts/whole, reductionism/holism, or particulars/universals can easily dissolve and reappear according to our approach. We can reduce reality and systems to our hearts' satisfaction, provided we remember where and when we started reducing from. We can take the pieces apart as far as we desire, provided we keep track and recall how to put them back together again. A fuller understanding will always come from such a two-way process.

Somehow, the duality game appears to follow certain common, almost universal pathways. Finite and infinite, for example, relate to conscious and unconscious. The same with unfolding and enfolding, or non-living and living, artificial or natural mechanisms and so on. Leibniz, for example, famously maintained that a living thing is a kind of divine automaton. What makes a divine automaton, he would profess, is the fact that "machines of nature, that is, living bodies, are still machines in their smallest parts, to infinity. It is this which constitutes the difference between nature and art, that is, between divine art and ours." From this realization, we have arrived to the current prevailing thesis, actually hubris, of "artificial intelligence" that we are all machines or that machines can be us, if not now then one day. To the delight of the strict reductionist molecular biologists this is a most welcome ally as it would render all life to be reduced to a big molecular machine. But what they forgot while descending blindly into their one-way dark alley is *infinity*. Infinity that cannot be ignored any more, for example in the newly fast developing field of quantum biology. There infinity sneaks in from the back door as quantum probabilities that necessarily have to take up the stage. Chasing the mind to the realm of the minuscule particles inside the brain, even Richard Feynman had to proclaim that: "Mind must be a sort of dynamical pattern, not so much founded in a neurological substrate as floating above it, independent of it."[21] But as far as patterns are concerned, "it takes one to recognize one." As patterns are an infinitude, it follows that it takes a Mind to see the pattern that is mind.

Left-Right Brain, Upside-Down Mind: Self-Locking and Self-Releasing Objectifications

> *Logic merely sanctions the conquests of the intuition.*
> — JACQUES HADAMARD

Iain McGilchrist, the author of the monumental work *The Master and his Emissary: The Divided Brain and the Making of the Western World*, takes the most up-to-date and deepest study on a different duality.[15] This time the duality concerns anatomical as well as functional aspects of the brain and their reflection of the dual aspects, logical-vs-creative, of the mind. He reevaluates, in a deeply erudite "tour de force," the seminal experiments of split-brain research since the late 1950s and the immense literature that has been generated ever since.

Usually the left hemisphere is attributed to the logical, analytic faculties of "how" the world is, where the right hemisphere is providing the relational, contextual meaning of "why." These two ways of understanding project two seemingly incompatible versions of the world, with quite different priorities and values. McGilchrist, unlike most scientists since the 1950s, does not abandon the attempt to understand why this division of the brain into two hemispheres is essential to human existence. In the course of his studies, he brings forth a very important realization: the duality is not so clear cut! He demonstrates, by a wide array of supporting experimental data and analyses, that every type of function is possible only due to the complementary concerted action of not only one but both of the hemispheres. The communication between the two major parts of the brain is what makes the whole brain serve its purpose for the benefit of the animal that possesses it. He goes on to argue that the differences lie not, as has been assumed so far, in the "what," but in the "how" the processes of each hemisphere play out their roles. More importantly, he emphasizes the not-so-well-known fact that the relationship between the two hemispheres is not symmetrical, either anatomically or functionally. To do that he utilizes in a wise way the perennial theme of the Emissary (the left hemisphere) and his Master (the right hemisphere). As the theme of the story goes, the master delegates, in good faith, valuable executive power to his emissary in order to carry out tasks that the right hemisphere cannot itself undertake. Yet, as the emissary has his own agenda, he can finally trap and betray its Master. The "How" becomes now more important and singularly so. The "Why" becomes secondary and of a lesser priority. Utility dominates value, usage oversteps beauty, the means disregard the ends. More or less these are the lines upon which the drama of our civilization unfolds. What the Emissary can never accept or realize is that by betraying the Master he also betrays himself.

The issues of brain plasticity, the complex role of the corpus callosum (the connecting bridge of the hemispheres), and the "inner sanctum" of the brain (the midbrain where the pineal gland and all the primitive functions reside) are also examined deeply and their interrelationships with the hemispheres are elucidated. The structure and function of the brain and its hemispheres are found to be determining factors, but not something that the mind can be reduced to. The hemispheres, we come to understand, are not mere machines with specific functions. These parts offer whole, self-consistent, versions of the world. If communication breakdown between the logical and relational parts persists, as our history of ideas testifies, in this "uneasy relationship" of theirs, we may unfortunately witness the final triumph of the left hemisphere: The Emissary's total control and the final dismissal of the Master, "at the expense of us all" as McGilchrist warns us.

Is it possible that these two modes of knowing and being, the analytical "left-brain" and the relational "right-brain" have their own reasoning?

But since the relational does not use any kind of representations—it only exists as sense and feeling—one expects the enterprise of deciphering it to be a futile, delusive one. To be more accurate, any attempt to analyze it in terms of logical structures would not reveal its "logic," it would just describe how the analytical logical structuring reflects on, and takes in, the relational. In other words, if there is any logic for the mostly unconscious processes of the relational "right brain" processes, it is a logic to the degree understood by the analytical, rational, mostly conscious "left brain."

Recently, with the proliferation and relative ease of formal logical systems implementations another body of work, undeservingly forgotten, emerges to the foreground. It is Matte Blanco's original theory and its latter variations, about the way the unconscious mind structures itself. The original work of Matte Blanco started to be known outside the academically closed psychoanalytic circles when he published his work "The Unconscious as Infinite Sets" around the end of the 1970s.[16] His discovery that emotions are "similar to mathematically infinite sets" opened new pathways and introduced a fresh new way of thinking about and in psychoanalysis. This work, which is difficult to find and follow, as it is quite esoteric yet not obscure, remained out of reach to wider audiences until the beginning of this century. With the renewed interest in non-classical logics, more and more introductory and explanatory texts emerge, using many vivid clinical examples and less complicated language.[13,17] So Matte Blanco's theories and ideas about "bi-logic" are re-emerging now, notably for their use in therapy. The "bi-logical" treatment asserts that there are two (hence the prefix "bi" in "bi-logic") distinct modes of reaching conclusions—one the conventional logico-analytical mode, and the other the unconventional felt-relational logic of the unconscious.

The conventional logic is asymmetrical while the unconventional one is symmetrical. In asymmetrical logic a proposition, A, is different from its negation, not-A. In the symmetrical logic, A and not-A are identical! Though Matte Blanco makes a delicate distinction between the Freudian sub-conscious and the unconscious, let us refrain from details of psychoanalytic terms. The important idea here is that analytical, left-brain, explicate order all share asymmetric logics, while the relational, right-brain, implicate order all share symmetric logics. Matte Blanco acknowledged that symmetric logic can be really upsetting to the logician and he had his share of polemics at the time. But the key is to understand that the symmetric logic works through *associations* rather than *propositions* and that it is *unquantified*. Properties have no degrees: they are all or nothing. It looks crazy but who said that the workings of the unconscious should look sane? Here is an example paraphrased from Mate Blanco's theory: The patient is bitten by a dog and hurries to the dentist. Why? Because the road of "explanatory deduction" revealed by the patient's prevailing symmetric logic is as follows: "– A dog has bitten me – It hurts – It is bad – The dog is bad – Dog's teeth are bad – I suffer from bad teeth – when suffering from bad teeth we go to the dentist"...QED!

As Eric Rayner, one of the key proponents of the theory and practice of Matte Blanco, puts it, "It illuminates the emotions behind thinking and the thoughts behind emotions."[17] The interplay between conscious and unconscious logics in bi-logic is for all those concerned with advanced psychoanalytic thinking and therapy, yet it sounds the familiar tone of the language of the Artist, the Poet and the Mystic. It has the trace and taste of "coinsidentia oppositorum" (coincidence of opposites). The avenues of research that it opens are very important to the study non-conventional logics. Especially for those logics where the observer creates context by his mere presence and choices for the observables. Oscar Wilde once said, through the mouth of the character Algernon Monchief in his theater play *The Importance of Being Earnest, A Trivial Comedy for Serious People*, (Act I), that "The pure and simple truth is rarely pure and never simple." ... So much even so, if the pure and simple truth is witnessed by observers with the faculty of introspection.

McGilchrist's and Matte Blanco's voices join many others' in a theme frequently encountered in our times: The prevailing strict, dogmatic adherence to one-way delineated thinking is as if we are happy to achieve record speed in a direction but we chose to ignore where it leads. Emilios Bouratinos also has been advocating a fundamental similar duality that leads us down the same path due to one-way of dogmatisms. He considers through his deep philosophical analysis how we appreciate, apprehend, and comprehend the world. His take on Consciousness, Objectivity, and Science illuminates each through the light of the others.[19] As he observes, the fundamental workings of Consciousness shape the world and also our ideas about the world. Here the one pole is what he calls "self-locking objectification" while the other is what he calls "self-releasing objectifica-tion." In between, or behind, this duality is the unifying substrate and at the same time the bridge in "Logos."

Logos originally meant ratio, proportion, analogy, as well as reason and relation, and all at the same time. As Bouratinos observes, reason or logos (originally derived from an understanding of the flexible rela-tionship of things to one another) became "formal logic" and "scientific methodology."[19,25] Formalism overtook intuition, prediction overshad-owed openness. Concepts became more rigid in the same way language adapted to this rigidity and perception filters selected and mapped every-thing in terms of objects, object-oriented, and object-mediated relations. This is the process of "self-locking" objectification where consciousness is trapped by its contents. It requires a return to a fresh communion with process, even in the sense of Whitehead, a daring opening to the forgot-ten flexibility of metaphor and analogy to unlock our consciousness from the grip of its own contents. This is the process of "self-releasing" objec-tification. Usually it comes with the familiar sense of the enlightenment due to context: by the descending of a vision of a "Why" on the "How" things are as they are. In science, this is related our familiar "Eureka!" moment. The moment of transcendence of an old paradigm by the dis-covery of a new one. Yet, this relation is just an instance of the faculty of our consciousness for "self-releasing" objectification. The overcoming of the need for paradigmatic thinking itself is the real boon of the inter-play between the "self-releasing"/"self-locking" game of objectification. This does not mean that we shall abandon paradigms altogether. To the contrary, we shall embrace them as what they really are: yet another em-issary of their master, Logos.

One might dismiss all this as just a romantic tendency or a return to philosophizing "as it used to be," before analytic philosophy and positiv-ism put philosophy in the straight-jacket of exact formalism and precise definitions. It might well be. But what is philosophy other than the love of wisdom? Asking Science to rediscover Her origins, the beauty and light that comes from an understanding of both how things are and why things are is, after all, primarily an act of Love. An act that the Master cannot delegate anymore to his Emissary. An act that can unlock the shackles we self-imposed on our consciousness. Since "Love is introduced without any parent at all" as Francis Bacon (one of the founding fathers of modern science) put it, reintroducing it in the kingdom of Science might be easier than it seems. Emilios Bouratinos advocates and leads this reintroduction through his approach. This enterprise he is inviting us into has to be equally precise as science in its means and joyful as art in its end: The regeneration of Science comes from a regeneration of Consciousness.

The suggestion he puts forward is an initiative to be undertaken for gradually creating a consciousness-informed science. As he says,

> ...for such an enterprise to succeed, it must be carried out in the light of the major scientific breakthroughs of the 20th century as they contain precious clues about how object-mediated thinking operates. Researchers and philosophers must be encouraged to examine their own personal understanding of how consciousness works and what are the conditions necessary for science to explore it.

An important input from 20th century Complexity Science, the understanding of the phenomenon of Self-organization, can secure the effectiveness of such questioning. The people and institutions involved in such an initiative should consider their task, assess their findings and grow organically in the light of inter-personal dialogue. Like David Bohm, who also advocated a form of dialogue that explores personal and group introspection, similar forms of dialogue, which Bouratinos calls "Inter-personal dialogue," can be found and harvested from "techniques practised by pre-literate societies throughout the globe before the advent of individualism. These techniques contain and channel 'the ego explosion' aiming at getting collectively to the bottom of any important issue to the 'tribe.' Inter-personal dialogue is effective as it creates a common conceptual and ontological ground among discussants." As he puts it, "How a conclusion is reached matters as much (if not more) as what the conclusion itself stipulates."

Evolving Science, Extending Science

Theories are nets: only he who casts will catch.
— NOVALIS

The subtlety of understanding depends on the understanding of subtlety.
— EMILIOS BOURATINOS

We have seen that even the idea of evolution has evolved and the prevailing "molecular machine" paradigm was exhausted by reaching its limits. A common theme for many areas of modern science, as well as modern thinking even in the more formal of formal sciences, is Logic. As we observed, logic has evolved to embrace realms of reasoning far and wide. From the principle of complementarity we learned that no matter how much a pole of a duality grows it cannot engulf the other pole in its entirety. Behind every complementary duality there is a coincidence of opposites lurking at its deepest level behind the complementary phenomena. Theory and experience notwithstanding, they also both point to a deeper level where their coexistence is based. We need theory to bring forth more data as we need more data to bring forth new theories.

De bètacanon by Fokke & Sukke
J. Reid, B. Geleijns , J-.M. van Tol

There is a Dutch comic strip (created by writer and illustrator Jean-Marc van Tol, and writers John Reid and Bastiaan Geleijnse called "Fokke & Sukke") where a dismissive, austere, professor exclaims when his student has just demonstrated an experiment: "Very impressive, dear colleague, but does it also work in theory?" As always, there will be facts and givens of experience that no theory can entirely explain. Denial is the first reaction and it is humanly so. The barrier for any understanding needs energy to overcome, it also needs patience and persistence. In contemporary cosmology, the issues of "dark matter and "dark energy" fuel fiery debates and open many discussions. But also in science at large it seems that we have a rapid accumulation of "dark matters" and "dark energies." Moreover, as the crises of our times keep dragging on we see an increasing polarization among cultures. The conflict between "the two cultures" (sciences and humanities) as famously delineated by C. P. Snow some decades ago, now has become a chaotic "meta-modern" battleground aiding a continuous proliferation of sub-cultures.[29] Numerous "mainstream" established fortifications prevent genuine dialogue, and on the other hand certain "new-age" misinformed groups create confusion about several very important issues. Signs and symptoms of a phase transition as they are, they nevertheless call for a deeper approach in thinking beyond mere paradigms. It is about time that we must concern ourselves not only with the study of nature but also by the nature of this study. Self-reflection and a quest for a new kind of validation of experience can be the only trusted peacemakers in resolving these contemporary conflicts.

We might be in just that instant of our collective evolution where we clearly see now the limitations of fortified self-interests and doctrinal ways of scientific thinking in society, the environment, the economy, politics, and education. The overarching theme in mainstream thinking is the seeking out of the "mechanism" as the core of any desired explanation. Although such mechanistic linear thinking ceased to be the prevailing one in physics since the beginning of the last century, other sciences are yet to catch up, still trying to fathom their practice in the mechanistic, naively reductionistic paradigm. They unquestioningly take their mode of understanding as only by means of reducing any operation to a mechanical process. They seek more and more the utility of the machine than the understanding of the process as a systemic whole. Hence crises ensue. And in our day when crises are met it is customary to throw up our hands and proclaim "Oh, this is complex" (end of discussion, thinking stops here!). I would propose instead to engage and encounter these complexities. Observe our limitations and navigate through them. Participate during our observations. Engage with systems and concepts. Be able to re-equip and re-inform our science by allowing it to reflect on its own foundations.

Arthur Koestler has remarked that the "decisive advances in the history of scientific thought can be described in terms of mental cross-fertilisation between different disciplines."[30] Complementary spirit is the key here. We shall be inspired by Socrates' "science of sciences." Such a "science of sciences" demands that we are not bound by paradigmatic thinking or doctrine. We must turn the investigative powers of "science-as-we-know-it" onto itself, then onto the scientists, and finally onto the major expressions of social life. This is the true meaning of the founding spirit of modern science, endangered by the one-way, solely utilitarian, version of science demanded today. Science as its prerequisite has the underlying principle of "Libre Examen" (The Freedom to Examine). This is a basic human right demanded by the early humanists for honest research and study, free of any chains that bind the mind. It remains our privilege today to turn this human right to our human responsibility. It is our responsibility and right to allow and to seek by reaching, and observing, our limits of investigation. Moreover, there is another dimension to it: by becoming aware of what limits our own thinking, we can become aware of what justifies the thinking of others. Moderation is not just a moral issue, it is what will reveal the ultimate complementarity of the opinions and methods of others. Inverting the parable: If we have a little mote in our eye, our neighbor can still see it clearly even if he has not cast the beam out his own eye! The complementarity of freedom and compassion always performed miracles—this time it can only be for the benefit of all.

Last but not least, we have to keep vividly in our mind that research is most useful when it is driven by the desire to satisfy curiosity. There is a tendency that when new knowledge is furnished there arises the need for its immediate applications. But the furnishing of new knowledge is not primarily due to the demand of applications. If Faraday and Maxwell were constrained to discover a better and stronger candle they would never discover electricity.[22] It took almost a century to fully illuminate and run our cities with electricity, but it would never have happened due to the results of applied candle science. We manage to discover important applications (the transistor, the laser, superconductivity) when we are driven by curiosity and desire to understand.[18] Creativity, like Evolution, is a playful activity. So this call for a new kind of science informed by a new kind of consciousness is also a call to bring back the fun in doing research for its own sake. This new science will offer themes for exploration and research not driven by an egocentric or institutionalized agenda, but from the mere pleasure of doing so for its own sake and making ends and means a complementary unity rather than an antagonistic duality.

Epilogue: A Modest Proposal

Thought must never be subjected either to a dogma,
to a party, to a passion, to an interest,
to a preconceived idea, or to anything.
For, to submit to it, would be to cease to be.
— HENRI POINCARÉ

The burden of industrial-driven research is drowning modern research. The diminishing support for basic research has already become a matter of concern not only for the academic world but also for societal bodies at large. More and more the old ways of supporting basic research are weakening, funds are disappearing, and young people are discouraged from pursuing research-related careers. If the dismissal of research and education as useless activities with no immediate profitable turnover continues at this rate, very soon they will cease to exist as realistic or even legitimate occupations.

Yet, as any crisis is pregnant with opportunities, novel organizational schemes appear strong and fast. Closing on a much more practical and utilitarian mode, it is worth adding that research and educational activities find new fertile and nourishing ground through the recent, and by now well established, activities of "crowd funding," "crowd sourcing," and what is called "participatory research." Interestingly, all these new forms of support relate to ideas from modern physics and complexity science, the first coming from the idea of self-organization in micro-economics while the second and third stem from self-organization in physics, biology, algorithmics, and distributed computing. They are all inspired, basically, by the self-organization and spontaneous division of labour theories dealing with hyper-organisms such as beehives and ant colonies. Actually, in recent years these novel fund-raising and resource-management ideas, operating via ad hoc assembled "crowds" interested in specific science projects small or large, drew the attention even of the "mainstream" scientific and research community, to the extent that the well-respected journals "Nature" and "Science" keep running special editorials to cover it.

Evidently, such actions liberate the scientific workforce from contractual, ordered research and the constraints of "Big-Science," "Big-Pharma," and other "Big-Money" strictly utilitarian guidelines. Most importantly, by actively engaging every interested party they promote and nurture, in the most efficient way, public awareness via public participation. One hopes that new creative forces also will be released towards a new kind of science as has been the case for any innovative breakthrough since the beginning.[18]

Moreover, and quite surprisingly, certain modern-day politicians, science policy advisers, activists, and CEOs use extensively the insights from Complexity Science, Chaos Theory, and Nonlinear Dynamics to elaborate on the theme of emergence as a new framework of understanding for the dynamics of "the workplace." One key issue that draws their attention is how coherence is established in an organism or organization, naturally born or human made. On the basis of organic coherence lies the phenomenon of "entrainment," where the alignment of a system's period and phase to the period and phase of one of its own subsystems is responsible for the emergence of collective modes of behaviour that might surpass and guide the dynamics of the whole organism. A coordinated small group of individuals will influence the motion of a much bigger crowd. A few coordinated and well-informed bee-scouts help the whole swarm to make decisions by safely reaching consensus. A system of interconnected hubs determines the robustness and effectiveness of the World Wide Web and many other networks. A lively interacting team of scholars can provide access to more knowledge than a whole group of universities.

Complexity science, like system science before it, has developed the necessary tools and concepts to deal with such emerging self-organization. The call, the imperative, is the formation of polycentric networks where projects and ideas are shared and circulated among a network of organizations, laboratories, and individuals. Self-organized and engaged in dialogue along a polycentric scheme, projects and ideas can be developed and are developing through such "commoners' science." Yet we cannot but notice the absence of cooperative research projects aimed at basic and applied research activities with a long-term horizon. I think it is within our reach to encourage and support the formation of multi-state and multi-stake cooperatives of individuals and labs for pure basic research within an interdisciplinary spirit and a self-reflecting attitude. Imagine this new kind of network as an evolving village, or even better, as an evolving organism like a "mycelium." Flexible, self-organized, exchanging energy, ideas, and nutrients with its environment. Open to societal changes and needs, yet resilient and growing, where it can grow, or keeping its ground and prepare to grow where it cannot grow. We can give birth to a live and resonant network of people, ideas, and projects. We can definitely envision it and organize it in such a fashion.

There is hope in considering an extended science, as sketched above. Contributing, cooperating, and sharing among numerous workers and thinkers, many unanswered questions will resurface allowing us to be able to move beyond accepted unquestioned answers. Many will be the questions, large and small, that will find new frameworks for investigation. For example, the question of information, memory, and knowledge-dynamics and their role in evolution, biological or not; or, what are the plausible frameworks where we can ask whether or not Nature has a mind of her own? What are the substrata that awareness/cognition/intelligence requires to express themselves, as Richard Feynman anticipated? Can there be any observables associated with it, and what type of observations can be expected? In what sense can these observables be measurable or felt? Or even; How to verify reality? Of course, the big questions about Consciousness will also ask for accommodation. And that is where all the difference will be made, and the greatest, deeply satisfactory and soul-nourishing fun will be found!

Endnotes

1 Gerald M. Rubin, "The draft sequences: Comparing species." *Nature* 409, 820-821 (15 February 2001).

2 Eugene V. Koonin, "Does the central dogma still stand?" *Biology Direct 2012*, 7:27.

3 Richard Strohman, "Epigenesis and Complexity: The Coming Kuhnian Revolution in Biology." *Nature Biotechnology*, Vol. 15, pp. 194-200, March 1997.

4 Alain E. Bussard, "A Scientific Revolution?" EMBO Rep. 2005 Aug; 6(8): 691–694. Science and Society, Viewpoint. (Alain E. Bussard is Honorary Professor at the Pasteur Institute, Paris, France.)

5 Charles H. Smith and George Beccaloni, eds., *Natural Selection and Beyond: The Intellectual Legacy of Alfred Russel Wallace*. Oxford University Press, 2008.

6 Henri Bergson, *Creative Evolution*. (1911), Dover Publications, 1998.

7 Abraham Pais, Niels Bohr's Times, In Physics, Philosophy and Polity. Oxford Clarendon Press. pp. 441-444, 1991.

8 V. Basios and E. Bouratinos, "Goedel's other legacy and the imperative of a self-reflective science." In Horizons of Truth: Goedel Centenary. University of Vienna 2006. Journal-ref: Kurt Goedel Society Collegium Logicum, vol. IX, pg. 1-5, 2006.

9 M. Katsumori, Niels Bohr's Complementarity: Its Structure, History, and Intersections with Hermeneutics and Deconstruction. Springer, Boston, (2011).

10 F. A. M. Frescura and B. J. Hiley, "Algebras, quantum theory and pre-space." published in Revista Brasileira de Fisica, Volume Especial, Julho 1984, Os 70 anos de Mario Schonberg, pp. 49–86, p.2

11 D. Bohm and B.J. Hiley, *The Undivided Universe: An ontological interpretation of quantum theory*. London, Routledge, 1993.

12 H. Gatti, *Giordano Bruno and Renaissance Science*. Cornell University Press, 1999.

13 C. Clarke, *Knowing, Doing and Being*. Imprint Academic, Exeter, 2013.

14 R.G. Jahn and B.J. Dunne, *Consciousness and the Source of Reality: The PEAR Odyssey*. ICRL Press, Princeton, NJ, 2011.

15 I. McGilchrist, *The Master and his Emissary: The Divided Brain and the Making of the Western World*. Yale University Press, 2009.

16 M Blanco and I. Duckworth, *The Unconscious as Infinite Sets*. London, 1975.

17 E. Rayner, *Unconscious Logic, An Introduction to Matte Blanco's Bi-Logic and its Uses*. First published in 1995 by Routledge, London; Taylor & Francis e-Library, 2006.

18 A. Flexner, *The Usefulness of Useless Knowledge*. First published in 1939, with a companion essay by Robert Dijkgraaf, Princeton University Press, 2017.

19 E. Bouratinos, *Science, Objectivity and Consciousness*. ICRL Press, Princeton, NJ, 2017 (in press).

20 J. A. Scott Kelso, *Dynamic Patterns: The Self-Organization of Brain and Behavior*. Complex Adaptive Systems, The MIT Press, 1995.

21 J. Gleick, *The Life and Science of Richard Feynman*. Pantheon, 1992.

22 B. Mahon, *The Man Who Changed Everything: The Life of James Clerk Maxwell*. John Wiley & Sons, 2004.

23 V. Basios, "Encountering Complexity: in Need for a Self-Reflecting (pre)epistemology." In *Endophysics, Time, Quantum and the Subjective*, edited by R. Buccheri, A.C. Elitzur and M. Saniga, World Scientific Press, 2005.

24 V. Basios, "Complexity, Interdependence & Objectification." In *Filters and Reflections: Perspectives on Reality*, edited by Z. Jones, B. Dunne, E. Hoeger, and R. Jahn, ICRL Press, Princeton, NJ, 2014.

25 E. Bouratinos, "A Science Toward the Limits." *Scientific & Medical Network*, 2003. https://old.scimednet.org/content/science-toward-limits (last accessed 20th of June 2017)

26 M. C. Boscá Díaz-Pintado, "Updating the wave-particle duality", 15th UK and European Meeting on the Foundations of Physics, Leeds, UK. 29–31 March 2007. http://philsci-archive.pitt.edu/3568/ (last accessd 20th of July 2017).

27 N. Bohr, "Light and Life," *Nature* 131, pp. 421-423, 1933.

28 H. Poincare, *Science and Hypothesis,* 1905, Dover abridged edition 1952.
29 C. P. Snow, *The Two Cultures*, Cambridge University Press, London (2001) [first print 1959].
30 A. Koestler, *The Act of Creation*, p. 230, Macmillan, New York (1964).

Contributors

Nelson Abreu co-founded the Institute of Applied Consciousness Technologies (I-ACT) in 2015 and has been a collaborator of non-profits IAC and ICRL since 2003. The Los Angeles-based electrical engineer specializes in power systems and applied consciousness with a special interest in consciousness technologies, innovation and conscious leadership. Nelson has practiced and trained others throughout North America and his native Portugal on phenomena related to chi and out-of-body experience since 1999. This is his fourth contribution to an anthology. He may be reached at nelson@conscioustech.com or nelson.abreu.ee@gmail.com

Wagner Alegretti is an electronic engineer who has worked with electricity generation, medical equipment and software development. He has been researching out-of-body experiences and developing means to detect life's subtle energy and to create a bioenergy technology since the 1980's. His most recent experiments use fMRI and other forms of nuclear magnetic resonance. Author of the book *Retrocognitions*, published in many languages, he is a co-founder of the International Academy of Consciousness and was its president from 2001 to 2014. He has presented at many events, including TEDx and the TSC (Tucson, USA). Email: wagner.alegretti@iacworld.org.

Thomas Orr Anderson is the founder of PhiSonics Sound Immersion Technologies and a long-time collaborator with ICRL and IAC. As a physicist, musician, inventor, and student-practitioner of a broad variety of natural healing arts, his endeavors are focused largely upon the utilization of sound and acoustics toward a broad variety of objectives. He has a undergone extensive training in a variety of contexts including four years as a private student of renowned holographer H. J. Caulfield and seven years as an apprentice of now 103-year-old Yaqui-Mayan medicine man, Tata Kachora. He may be contacted at thomas@PhiSonics.com.

Vasilieios Basios, Ph.D., is a senior researcher at the Physics of Complex Systems Department of the University of Brussels, conducting interdisciplinary research on self-organization, collective phenomena and emergence in complex matter as well as aspects of the foundations of complex systems. During his formative years he worked within the team of Ilya Prigogine at the Solvay Institutes for Physics and Chemistry in Brussels. He is interested on the history of ideas in science and their role in the transformation of science beyond the prevailing mechanistic world-view. Email: vbasios@gmail.com.

William F. Bengston, Ph.D., is a professor of statistics and research methods and President of the Society for Scientific Exploration. He has been researching anomalous healing for over thirty-five years, and has produced the first successful full cures of transplanted mammary cancer and methylcholanthrene-induced sarcomas in mice by energy healing techniques that he helped to develop. He has numerous publications in scientific journals and has lectured widely in the U.S., Canada, and Europe. His memoir, *The Energy Cure*, is published by Sounds True. Email: wbengston@sjcny.edu.

Richard A. Blasband, M.D., worked for over 50 years as a Psychiatric Orgone Therapist, conducted original research in the fields of orgone biophysics and meteorology, and published over 60 papers in this field and on consciousness. He currently lives in France and works as a healer using the Levashov Method of Intentional Healing. He may be contacted at richardablasband@gmail.com.

Ulisse Di Corpo and Antonella Vaninni, Ph.D., are co-authors of *Syntropy: The Spirit of Love,* as well as several other books and numerous publications. Ulisse holds a Ph.D. in statistics and social research and works in the field of social research, providing methodological support and software tools to researchers. Antonella is a practicing psychotherapist and hypnotherapist with a PhD in cognitive psychology, and is a Tai Chi instructor and expert in martial arts. Together they are developing the vital needs theory, first proposed by Ulisse in 1977, and the uses of the law of syntropy in the field of psychotherapy. Email: ulisse.dicorpo@syntropy.org.

Steve Curtis is a visionary leader and venture capitalist whose mission is to elevate human consciousness as a call to spiritual reawakening. His business interests include healing food and drink, biofeedback technology, and ethnobotanicals. From a diagnosis of a terminal cancer at twenty-four with two years to live, he mastered his health by prioritizing healthy eating, love, and mastering fear – including a summit up Mount Everest. His extraordinary life has led him to teach the rest of the world out of illness through his nonprofit, Perception Medicine Foundation, dedicated to broadening our understanding of how the human mind directly influences disease. Email: miacosco@gmail.com.

Larry Dossey, M.D. is a distinguished physician and internationally influential advocate of the role of the mind in health and the role of spirituality in healthcare. He served as a battalion surgeon in Vietnam, where he was decorated for valor, and was Chief of Staff of Medical City Dallas Hospital. He is the author of twelve books and numerous articles and is Executive Editor of the peer-reviewed journal *Explore: The Journal of Science and Healing.* He has lectured all over the world, including many major medical schools and hospitals in the United States. Email: larry@dosseydossey.com.

Brenda Dunne and Robert Jahn, Ph.D., were Director and Laboratory Manager, respectively, of the Princeton Engineering Anomalies Research (PEAR) laboratory in Princeton University's School of Engineering and Applied Science from its inception in 1979 until its closing in 2007. Together and individually they have authored five books and numerous technical publications. Bob is a retired professor of aerospace engineering and recipient of the prestigious AIAA Wyld Propulsion Award, among other notable honors, and is Dean Emeritus of Princeton's engineering school. They currently serve as Chairman and President/Treasurer of International Consciousness Research Laboratories (ICRL) and the ICRL Press. Email: bjd@icrl.org.

Antonio Giuditta, Ph.D., is a Professor of Department of Biological Sciences, at the University of Naples, who studies basic concepts of neurobiology regarding the local system of gene expression in axons and synaptic regions, the role of sleep in memory processing, the role of brain DNA in brain plasticity, and theoretical problems related to the nature, philogenetic origin and role of mind, and to the mechanism of biological evolution. Email: giuditta@unina.it

Brennan Kersgaard is a young scholar, researcher, and artist residing in Boulder, Colorado, with an interest in the intersection between ancient spiritual practice and modern scientific methodology. Specifically, his research focuses on the relationship between light and life and the neurobiological underpinnings of mystical states of consciousness. He is a graduate of the University of Colorado, former ICRL intern, and published author in the field of psychopharmacology and mind-altering drugs. When not studying, you can find him in the mountains, on his yoga mat, or making art where he loves to transmute his intellectual research into his direct life and experience for his own spiritual exploration and growth. Email: brennankersgaard@gmail.com.

Rollin McCraty, Ph.D., is Director of Research at the Institute of HeartMath in Boulder Creek, CA. His research interests include the physiology of emotion, intuition and optimal functioning. His work focuses on the mechanisms by which emotions influence cognitive processes, behavior, and health as well as the global interconnectivity between people and the earth's energetic systems. Email: rollin@heartmath.org

Rupert Sheldrake, Ph.D., is a biologist and author of more than 85 scientific papers and 11 books, including Science Set Free (called The Science Delusion in the UK). He was a Fellow of Clare College, Cambridge, a Research Fellow of the Royal Society, and Director of the Perrott-Warrick Project, funded from Trinity College, Cambridge. He is a Fellow of Schumacher College, Dartington, England, and the Institute of Noetic Sciences, Pertaluma, CA. He lives in London. Web site: www.sheldrake.org

Yoléne Thomas, Ph.D. After a medical cursus, Yoléne Thomas received her Ph.D. in immunology from the University of Paris. In 1979, she joined the University of Columbia Medical School first as a postdoctoral fellow and then, as an assistant professor. In 1987, she returned to France and became a research director at the CNRS. She joined the laboratory of Jacques Benveniste to set up her own immunology team and had the good fortune of being able to collaborate with him for over 16 years. Email: yolene@noos.fr.

www.ingramcontent.com/pod-product-compliance
Lightning Source LLC
Chambersburg PA
CBHW070247200326
41518CB00010B/1714